MIX
Papier aus verantwortungsvollen Quellen
Paper from responsible sources
FSC® C105338

Pritesh Ranjan Dash

Phytochemical Screening and Pharmacological Investigations on *Hedychium coronarium*

Anchor Academic
Publishing

Dash, Pritesh Ranjan: Phytochemical Screening and Pharmacological Investigations on *Hedychium coronarium*, Hamburg, Anchor Academic Publishing 2016

Buch-ISBN: 978-3-96067-036-0
PDF-eBook-ISBN: 978-3-96067-536-5
Druck/Herstellung: Anchor Academic Publishing, Hamburg, 2016

Bibliografische Information der Deutschen Nationalbibliothek:
Die Deutsche Nationalbibliothek verzeichnet diese Publikation in der Deutschen Nationalbibliografie; detaillierte bibliografische Daten sind im Internet über http://dnb.d-nb.de abrufbar.

Bibliographical Information of the German National Library:
The German National Library lists this publication in the German National Bibliography. Detailed bibliographic data can be found at: http://dnb.d-nb.de

All rights reserved. This publication may not be reproduced, stored in a retrieval system or transmitted, in any form or by any means, electronic, mechanical, photocopying, recording or otherwise, without the prior permission of the publishers.

Das Werk einschließlich aller seiner Teile ist urheberrechtlich geschützt. Jede Verwertung außerhalb der Grenzen des Urheberrechtsgesetzes ist ohne Zustimmung des Verlages unzulässig und strafbar. Dies gilt insbesondere für Vervielfältigungen, Übersetzungen, Mikroverfilmungen und die Einspeicherung und Bearbeitung in elektronischen Systemen.

Die Wiedergabe von Gebrauchsnamen, Handelsnamen, Warenbezeichnungen usw. in diesem Werk berechtigt auch ohne besondere Kennzeichnung nicht zu der Annahme, dass solche Namen im Sinne der Warenzeichen- und Markenschutz-Gesetzgebung als frei zu betrachten wären und daher von jedermann benutzt werden dürften.

Die Informationen in diesem Werk wurden mit Sorgfalt erarbeitet. Dennoch können Fehler nicht vollständig ausgeschlossen werden und die Diplomica Verlag GmbH, die Autoren oder Übersetzer übernehmen keine juristische Verantwortung oder irgendeine Haftung für evtl. verbliebene fehlerhafte Angaben und deren Folgen.

Alle Rechte vorbehalten

© Anchor Academic Publishing, Imprint der Diplomica Verlag GmbH
Hermannstal 119k, 22119 Hamburg
http://www.diplomica-verlag.de, Hamburg 2016
Printed in Germany

Dedicated
To My
Best Friend
Pia

Publications

1. Evaluation of Analgesic and Neuropharmacological activities of methanolic rhizome extract of *Hedychium coronarium*. *International Journal of Pharmaceutical Sciences and Research*, 2011, **2**(4): 979-984.

2. Preliminary studies on phytochemicals and cytotoxic activity of methanolic rhizome extract of *Hedychium coronarium*. *Journal of Pharmacognosy and Phytochemistry, 2015,* 4(1): 136-139.

ACKNOWLEDGEMENT

At the very beginning, all gratefulness to Almighty creator "God" who has enabled me to complete this research, manages each and everything soundly.

I am greatly indebted to my honorable supervisor, **Moni Rani Saha**, Assistant Professor, Department of Pharmacy, Stamford University, for her expert supervision, suggestions, untiring help, guidance, and enthusiastic encouragement throughout the entire period of the research. Without her supervisions my research would have never been accomplished.

I would like to express deep gratitude and warm regards to Professor **Dr. Abdul Ghani**, Former Chairman, Department of Pharmacy, Stamford University Bangladesh for his heartiest co-operation and good wishes.

My sincere gratitude goes to all of my teachers of the Department of Pharmacy, Stamford University Bangladesh, for extending their helping hands and affectionate attitude whenever I needed.

Finally, I would like to express my sincere gratitude to my best friend Pia and my Parents for their love, hearty encouragement and unselfish sacrifice during my study.

I truly believe that all people whom I have not personally mentioned here are aware of my deep appreciation

January, 2016

Author
Pritesh Ranjan Dash

CONTENTS

Serial No.	Title	Page no.
Abstract		11
CHAPTER 1	**INTRODUCTION**	12-22
1.1	General Introduction	12-15
1.2	Medicinal plants	15
1.3	The history of medicinal plants	16
1.4	Contribution of medicinal plants of modern medicine	16-17
1.5	Prospect of herbal drug research in Bangladesh	17-18
1.6	Economic significance of medicinal plants	18-19
1.7	Crude drug	19
1.8	Rationale of the present work	20-21
1.9	Aim of the present work	22
1.10	Plan of the present work	22
CHAPTER 2	**LITERATURE REVIEW**	23-31
2.1	Plant review	23-28
2.2	Phytochemical properties	28-30
2.3	Pharmacological properties	31
CHAPTER 3	**MATERIALS AND METHODS**	32-64
3.1	Plant material	32-34
3.2	Materials used in the study	35-37
3.3	Phytochemical screening	38-43
	Result and discussion	43
3.4	Pharmacological investigations	44-64
3.4.1	Analgesic activity test	44-52
	Acetic acid induced writhing test	44-48
	Result	49
	Tail immersion test	49-50
	Result	51
	Discussion	52
3.4.2	Neuropharmacological study	53-58
	Hole cross method	54-55

Serial No.	Title	Page no.
	CONTENTS	
Serial No.	**Title**	**Page no.**
	Result	55
	Open field method	55-56
	Result	56-57
	Discussion	57-58
3.4.3	Cytotoxic activity test	58-64
	Brine shrimp lethality bioassay	58-64
	Results and discussion	61-64
CHAPTER 4	**CONCLUDING REMARKS**	65
CHAPTER 5	**REFERENCES**	66-72

LIST OF TABLES

Table	Title	Page no.
1.1	Examples of Crude Drugs and Their Therapeutic Uses	20
3.1	Accession Code of the Plant	32
3.2	List of Apparatus Used for the Experiment	35
3.3	List of Glassware Used	35
3.4	List of Reagents Used for the Experiment	36
3.5	Detail information of the mice used for the Experiment	36
3.6	Type of Food Used for the Mice	37
3.7	Materials Used for Animal House	37
3.8	Reagents used for different group tests	38-39
3.9	Results of phytochemical screening	43
3.10	Experiment Profile to assess the effect of crude extract of *Hedychium coronarium* on acetic acid induced writhing of mice	47
3.11	Effect of methanol extract of *Hedychium coronarium* on acetic acid induced writhing test in mice	49
3.12	Effects of the methanolic extract of *Hedychium coronarium* on tail immersion test	51
3.13	Experiment Profile to assess the effect of crude extract of *Hedychium coronarium* on CNS depressant activity test on Mice	53
3.14	Effect of methanol extract of *Hedychium coronarium* on Hole cross test in mice	55
3.15	Effects of methanol extract of *Hedychium coronarium* on open field test in mice	56
3.16	Effect of *Hedychium coronarium* on brine shrimp lethality test in *Artemia salina*	61
3.17	Result of *Hedychium coronarium* against on *Artemia salina*	62

LIST OF FIGURES

Figure	Title	Page no.
2.1	Plant of *Hedychium coronarium*	26
2.2	Leaves of *Hedychium coronarium*	26
2.3	Rhizomes of *Hedychium coronarium*	27
2.4	White color flower of *Hedychium coronarium*	27
2.5	Whole part of *Hedychium coronarium*	28
2.6	Structure of chemical compounds isolated from *Hedychium coronarium*	29-30
3.1	Grinding by Using a Blender	33
3.2	Rhizomes extract of *Hedychium coronarium*	33
3.3	Extraction of *Hedychium coronarium* by maceration	34
3.4	*Swiss albino* mouse	36
3.5	Identification of experimental animals	37
3.6	Stock solution used for chemical group tests	42
3.7	Schematic representation of acetic acid induced writhing of mice for investigation of analgesic activity	45
3.8	Preparing mice for writhing by injecting acetic acid	46
3.9	Synthesis of Prostaglandins and Leukotrienes	47
3.10	Half Writhing Given by Mice	48
3.11	Full Writhing Given by Mice	48
3.12	Percentage inhibition of writhing reflex by *Hedychium coronarium*	49
3.13	Tail immersion test on mice	50
3.14	Percentage of elongation by *Hedychium coronarium* in tail immersion method	51
3.15	Hole Cross Test	54
3.16	Percentage of movements inhibitions by *Hedychium coronarium* in hole cross method	55
3.17	Percentage of movements inhibitions by *Hedychium coronarium* in open field method	57
3.18	Hatching of Brine Shrimps	60
3.19	Effect of methanolic extract of *Hedychium coronarium* on Brine shrimp nauplii.	62
3.20	Effect of Vincristine Sulphate *on* Brine shrimp nauplii	63

ABSTRACT

The present study was carried out for phytochemical screening and pharmacological investigations on methanolic extract of rhizomes of *Hedychium coronarium* (Local name: Dolan Champa, Family: Zingiberaceae). In this study, the possible analgesic, CNS (Central Nervous System) depressant activities of the methanolic rhizome extract of *Hedychium coronarium* were investigated at the doses of 100 mg/Kg, 200 mg/kg and 400 mg/Kg body weight on mice by oral administration. The analgesic activities were investigated for its central and peripheral pharmacological actions using tail immersion test and acetic acid-induced writhing test respectively. Its CNS depressant activity was evaluated by using hole cross and open field tests and the cytotoxic activity was observed using brine shrimp lethality bioassay. The result of preliminary phytochemical screening reveals that the methanolic rhizome extract contains alkaloids, flavonoids, saponins, carbohydrates and steroids. The result of Acetic acid induced writhing test reveals that the extract inhibited 26.15%, 47.94% and 73.12% of writhing at the doses of 100 mg/kg, 200 mg/kg and 400 mg/Kg body weight respectively, whereas writhing inhibition of the standard drug Diclofenac-Na was 73.36% at 25 mg/Kg body weight. On the other hand, the result of tail flick test also shows that the potential analgesic activity of the extract which is also comparable to the standard drug Diclofenac-Na (25 mg/kg body weight). The results of CNS depressant activity show that the extract decreased dose dependent motor activity and exploratory behavior of mice in both hole cross and open field test. The number of hole crossed in hole cross test and field crossed in field cross test decreased as time approached. The result of brine shrimp lethality bioassay reveals that the extract has dose dependent cytotoxic activity. The LC_{50} value of the methanolic rhizome extract was 0.39 µg/ml, whereas the LC_{50} of the reference anticancer drug vincristine sulphate was 0.52 µg/ml. These results suggest that the extract possesses analgesic and CNS depressant activity on mice and also has brine shrimp cytotoxic activity.

Key words: *Hedychium coronarium,* analgesic activity, neuropharmacological activity, phytochemical screening, cytototoxic activity.

CHAPTER 1: INTRODUCTION

1. 1 General Introduction

Plants and man are inseparable, because plants not only provide man with food, shelter and medicine, but also the life sustaining oxygen gas. Since disease, decay and death have always co-existed with life, the early man had to think about disease and its treatment at the dawn of human intellect. (Kirtikar and Basu, 1980). Thus, the human race started using plants as a means of treatment of disease and injuries from the very early days of civilization on earth and in its long journey from ancient time to the modern age it has successfully used plants and plants products as effective therapeutic tools for fighting disease and other health hazards.

As therapeutic use of plants continued with the progress of civilization and development of human knowledge, scientists endeavored to isolate different chemical constituents from plants, put them to biological and pharmacological tests and identify therapeutically active natural compounds, which have eventually used to prepare modern medicines. In course of time synthetic analogues and derivatives of the natural compounds were also prepared. In this way, ancient uses of Datura plants have led to the isolation of hyoscinic, hyoscyamine, atropine and tigloidine, Cinchona barks to quinine and quinidine, *Rauwolfia serpentina* root to reserpine and rescinnamine, *Digitalis purpurea* to digitoxin and digoxin; Opium to morphine and codeine, Ergot to ergotamine and ergametrine, Sena to sennosides, *Catharanthus roseus* to vincristine and vinblastine to mention a few. (Ghani, 1998).

In addition to these, there are many other plant-derived chemical substance of known structures that are used as drugs or necessary components of many modern medicinal preparations. These include camphor, capsaicin, eucalyptol, menthol, minor cardiac glycosides, various volatile oils, etc.

Facilitated by the rapid development of technology of isolation and characterization process, particularly chromatography and spectroscopic methods a large number of therapeutically active plant constituents have been isolate during the last two decades. Simultaneous advancement in the fields of medical, botany, chemistry, biochemistry, pharmacognosy and pharmacology has tremendously helped the discovery, isolation, characterization, structure elucidation and synthesis of new drugs plants, with the development of further newer techniques and methods of plant analysis and with the tremendous increase in man's knowledge of chemistry and pharmacology more and more medicinal compounds are likely to be discovered from plants.

Although with the advent of synthetic drugs the use and procurement of plant-derived drugs have declined to large extent, a large number of drugs of modern medicine are still obtained from plant sources. As more data become available from phytochemical analysis and pharmacological screening of medicinal plants, the number of plant-derived drugs of modern medicine is increasing everyday.

How prominently plant-derived drugs still feature in modern medicine can be assessed from the following facts (Ghani, 1998):

1. A recent survey by the United Nations Commissions for Trade and Development (UNCTAD) indicated that about 33% of drugs, produced in the development countries, are derived from plants (UNCTAD/GATT 1974) and that if microbes are added 60% of medicinal products are of natural origin. (Sofowara, 1982).
2. According to some sources almost 80% of present-day of medicines are directly or indirectly derived from plants (Mayers, 1892).
3. More than 47% of all drugs used in Russia, are obtained from botanical Sources (Ampofo, 1979).
4. From *Stephania cepharantha and Stephania sasabi* (Jap. Journ. Exp. Med. 1949, 1:69). In the United States, in 1980 alone, the consumer paid 8 billion dollars for prescription drugs in which the active ingredients are still alive from plants (Sofowara, 1982).

5. 47% of some 300 million new prescriptions written by physicians in America in 1961, contained as one as more active ingredients, a drug of natural origin (Fams-worth, 1966).

6. In 1960 47% of drugs, prescribed by physicians in the United States, of America, were from natural sources (Bingel *et al,* 1660).

7. In 1967 25% of the products, which appeared in 1.05 billion prescriptions filled in the United States, contained one or more ingredients derived from higher plants. (Karolkovans *et al*, 1966).

8. Even today 80% of the rural population of most developing countries of the world depends as herbal medicine for maintaining its health and well being (Ghani, 1987).

9. The consumption of medicinal plants in increasing in many developed countries, where 35% of drugs contain active principles from natural origin (Irvine, 1995).

10. The North America used 170 drugs from different plants, which are as official in the USP or NF. Surprisingly this large quantity of modern drugs comes from less than 15% of the plants, which are known to have been investigated pharmacologically, out of an estimated 250000 to 500000 species of higher plants growing on earth (Farmsworth *et al*, 1985).

At present, thousands of plant metabolites are being successfully used in the Treatment of variety of diseases (Farmsworth *et al,* 1985). A few striking Examples of plant metabolites include taxol from *Taxus brevifalia* (Kumar *et al*, 1994), vincristine and vinblastine from *Vinca roseus* (Staskar, 1980), of which are important anticancer agents being used clinically. In the current popular field of chemotherapy, cepharanthine, isolated) is being used as a prophylactic in the management of tuberculosis.

In china, about 15,000 factories are involved in producing herbal drugs, herbal medicines have been developed to a remarkable standard by applying modern scientific technology in many countries such as, China, India, Bangladesh, Sri Lanka, Thailand and United Kingdom. In these countries, the dependence on allopathic drug has been decreased to greater extent (Borin, 1998) "Modern medicine still has much to learn from the collector of herbs" said Dr.Hafdan Mohler, Director General of World Health Organization. Many of the plants, familiar to the witch doctor really do have the healing power that tradition attaches to them. The age-old art of the herbalist must be tapped (Mayers, 1982).

Thus it is apparent that whatever progress, science might have made in the field of medicine over the years, plants still remain as the primary source of supply of many important drugs used in modern medicine. Indeed the potential of obtaining new drugs from plant sources is so great that thousands of substances of plant origin are now being studied for activity against such formidable foes as heart disease, cancer and AIDS,(Ogden et al, 1981).This type of studies are sure to bring fruitful results, because of the fact that plant kingdom represents a virtually untapped reservoirs of new chemical compounds, many extraordinarily, dynamic some providing novel bases on which the synthetic chemist may build even more interesting structures (Said, 1995). In this way, modern medicine will continue to be enriched by the introduction of newer and more potent drugs from plant sources.

1.2 Medicinal Plants

In nature plants of several variations are available which are responsible for various pharmacological actions. They are termed as medicinal plants. On the other hand, some of them produce harmful effects on animal system; they are termed as toxic or poisonous plants. It has now been established that the plants which naturally synthesize and accumulate some secondary metabolites like alkaloids, glycosides, tannins, volatile oil and contain minerals and vitamins posses medicinal properties (Ghani, 1998).A medicinal plant may thus be defined as a plant which, in one or more of its organs contains substances that can be used for therapeutic uses or which are precursors for the synthesis of useful drugs.

However, ideally a definition of medicinal plants should include the following: (Sofowora, 1982)
a) Plants or plants parts medicinally in galenical preparation (e.g. decoction, infusion, etc.)
b) Plants used for extraction of pure substances either for direct medicinal use or for the synthesis of medicinal compounds (e.g. synthesis of sex hormones).
c) Food, spice and perfumery plants used medicinally.
d) Microscopic plants, e.g. fungi, actinomycets, used for isolation of drugs, especially antibiotics.
e) Fiber plants, e.g. cotton, flax, jute, used for the preparation of surgical dressings.

A large number of plants are used as medicinal agents in this world. Specifically in Bangladesh about four hundred fifty species are used as medicinal plants (Ghani, 1998).

1.3 The history of medicinal plants

According to recorded history of human civilization, man was well aware of the medicinal properties of many plants growing around him more than five thousand years ago. The earliest mention of the medicinal use of plants is found in the Rig Veda (4500-1600BC), which reported that the Indo-Aryans used the Soma plant as a medicinal agent (Chopra et al., 1982).A large number of plants now known to contain important medicinal agents were known to contain important medicinal agents were known to the Babylonians (Ca 3000BC).The famous Papyrus Ebers recorded the use of many plants which were used by the ancient Egyptians as far back as 1500 BC Ayurvedic system of medicine described the use of 127 plants as curative agents as early as in 1200 BC.The first Chinese Pharmacopoeia containing a list of 135 different plant medicines with their use and methods of preparation appeared around 1122BC. More than 400 recipes of plant medicines used in the Greek system of medicine were described by Hypocrites around 400BC (Le Strange, 1977). Arabian Muslim Physicians like Al-Razi and Ibne Sina (9th to 12th century AD), brought out revolution in the history of medicine by bringing new drugs of plant and mineral origin into general use. The use of medicinal plants in Europe in the 13th and 14th centuries was based on the "Doctrine of Signatures" or developed by Paraclsus (1400-1541 AD), a Swiss Al-Chemist and Physician (Ghani, 1998). According to the superstitious doctrine all plants possessed some sign, given by creator, which indicated the use, for which they were intended. Thus a plant with heart shaped leaves should be used for heart ailments, the sap of blood root (Sanguinary condenses). As a blood tonic, the liver leaf with its three lobed for liver troubles, the walnut with numerous investigations and convolutions for brain diseases and pomegranate seeds for dental diseases and so on. (Graves, 1990).

1.4 Contribution of medicinal Plants to modern medicine

The term medicine refers to a preparation or compound containing one or more drug (s) or therapeutic agent (s) which is used in the treatment, cure and mitigation of various diseases and internal injuries of man and other animals. The preparation may also contain substances other than the drug (s). The drug (s) in the medicine is (are) the active therapeutic agent (s) that cures (cure) the disease or heals (heal) the wound or injury. "Modern medicines" are those medicinal preparations which are produced scientifically by using modern echnology and know- how and which are in current use for the cure and management of diseases.Modification, improvements, sophistifications and newer discoveries are continuously changing the type, quality, of medicinal preparations. The real changes that have taking place in medicines are those in their active ingredients, excipients and form of presentation. And these are the areas where plants have been contributing enormously since the human race first discovered medicine to ensure its existence on earth. (Ghani, 1998). Isolation of the natural analgesic drug morphine from the latex of *Papaver somniferum* capsules (Opium) in 1806 is probably the first moat important example of natural drugs which plants have directly contributed to modern medicine .Isolation of other important plant –derived drugs of modern medicine rapidly followed and may useful drugs have since been discovered and introduced into modern medicine. Drugs like Strychnine from *Strychnos nux vomica,* (1817), emetine from *Cephaelis ipecauanha* (1817), caffeine from *Thea sinensis* (1819), quinine from Cinchona spp.(1820) and colchicine from *Colchicium autumnal* (1820) constitute some examples of such early drugs. The list of the plant-derived medicinal in modern medicine is very big now. About 100 plant derived drugs of defined structures are in common use today throughout the world and about half of them are accepted as useful drugs in the industrialized countries. These include drugs like atropine, colchicine, deserpidine, digitoxin, L-lopa, emetine, ephedrine, ergometrine, ergotamine, hyoscine, hyoscyamine, physostigmine, picrotoxin, pilocarpine, pseudoephedrine, quinines, quinine, rescinnamine, reserpine, sennosides, sparetine strophanthin, strychnine, theobromine, theophylline, vinblastine, vincristine etc. Other plant derived drugs which are used widely but not generally in Western modern medicine include anabasine, andrographolide, arecoline, berberine, brucine, cannibal, cephaeline, cocanine, curcumin, glycyrrhizin, hesperidin, hydrazine, nicotine, palmitine, quercetin, rutin santonin, vincamine, yohimlin etc.(Farnsworth and Moris,1976; Farnsworth and Bingel, 1977).

1.5 Prospect of herbal drug research in Bangladesh

Approximately 80% of the world's population exclusively uses plants for various healing purposes. In the industrially developed countries almost 35% of the drugs contain active principles of natural origin and consumption of medicinal plants is increasing. (Irvinne, 1995). The practice of traditional medicine in China is firmly established. More than 5000 kinds of Chinese medicinal herbs are used medicinally. India and Thailand are two burning examples of such countries in this subcontinent which earn cores of rupees by exporting medicinal plants and their semi-processed products to other countries including Bangladesh. There are still others such as china, India and Pakistan which utilize their own medicinal plants for local manufacture of both. Eastern and Western medicines and pharmaceutical products (Ghani, 1998). The suitable weather and fertile soil have made Bangladesh a great source of medicinal plants .About five thousands herbs, shrubs, trees, aromatic and aquatic plants are scattered throughout the forests, jungles, hills, cornfields, plain field, roadsides, gardens, marshy lands and watery place of Bangladesh. These medicinal plants which contain steroids, alkaloids, tannins, saponins, glycosides acetogenin and resin are most useful medicinal agents. The socio-economic condition has far-reaching effects on the health condition of the Bangladesh population. Existing health care system is not very impressive. Though there is steady increase of facilities, the situation has not improved much. Till now only 30% of the entire population has been brought under primary Health Care (PHC) (Ali *et al.*, 1990). 20% of the populations of our country have access to the Western medicines and rest 75-80% of the rural population still receives healthcare services from the indigenous traditional Ethno medicine practitioners. They play a significant role in providing primary health care services to rural people. They serve as important therapeutic agents as well as imported raw materials for the manufacture of traditional and modern medicines. Substantial amount of foreign exchange can be earned by exporting medicinal plants to other countries (Ghani, 1998). From a survey in different villages of Bangladesh (Ahmed,1990),it has been seen that if people suffer from illness approximately 14% of them go the qualified allopathic doctors, 29% contact unqualified village doctors, 10% contacts mollahs, 29% approach quack and 19% contact homeopaths. The survey represents an extensive use of medicinal plants, most of which are served in a crude and substandard form by the different types of traditional practitioners which is sometime hazardous for the health. For that, we should standardize our traditional use of the medicinal plants. Thus to maintain a safer traditional practices we should make more research with our medicinal plants to determine their chemical entities and biological activities property.

1.6 Economic significance of medicinal plants

Medicinal plants are rich sources of bioactive compounds and thus serve as important raw materials for drug production. They may constitute a valuable natural asset of a country and contribute a great deal to its health care systems. Judicious and scientific exploitation of this wealth can significantly improve the general health of the people. Being a valuable commodity of commerce, a country can also earn a food amount of foreign currency by exporting this natural wealth to other countries. The Government of Bangladesh officially recognizes Unani and Ayurvedic system of medicine. They are dispensed as broken pieces, coarse and fine powders, pills of different sizes, in the form of compressed tablets, as liquid preparations, as semisolid masses and in the form of ointment and creams, neatly packed in appropriate sachets, packets, aluminum foils, plastics or metallic containers and glass bottles. The containers are fully labeled with indications, contaminations, doses and directions for uses and storage, just like modern medicine. If we could make proper use of our medicinal plants we could get medicines in low costs and then it might be possible to fulfill the demand of our medication. Thus supplying low cost medicines to our population we could establish a better health care system. And in order to achieve this goal, research and development on the traditional medicines should be given in the proper privilege. The leaves, flowers, fruits and roots are extensively used for treating cold, cough, whooping cough and chronic bronchitis and asthma, as sedative-expectorant antispasmodic and as anthelmintic. The drug is employed in different form, such as fresh juice, decoction, infusion and powder. Also was given as alcoholic extract and liquid extract or syrup. The dried leaf is smoked as a cigarette. It is also given along with other expectorants and forms a part of several proprietary compounds. In chronic bronchitis, it is efficacious and affords relief, especially when the sputum is thick and tenacious acting very much similar to ipecacuanha. The cough is relieved and the sputum is liquefied and easily expelled. The leaf juice is stated to cure diarrhea, dysentery.

1.7 Crude Drug

The substance of natural origin both from plant and animal source possesses therapeutic properties and pharmacological actions. These substances in the natural state comprise whole plants, their morphological or anatomical parts, saps, secretions etc. whole animals, their anatomical parts, glands or other organs, extracts, secretions of their organs. These drugs are used as therapeutic agents in many traditional medicinal preparations in everywhere (Ghani, 1998).A crude drug is a natural drug of plant, animal or mineral origin which has undergone no treatment other than collection and drying, that is the quality or appearance of the drug has not been advanced or improved by any physical or chemical treatment (Ghani, 1998).In Bangladesh numerous crude drugs are prepared from local plant and animal sources while many more are imported from foreign countries for use in the preparation of Unani, Ayurvedic and Homeopathic medicines. Many of them are also used in Hekimi, Kaviraji and Folk medicine practices in the country. Some of the official crude drugs available in Bangladesh are Abroma bark, Acacia, Aloes, Amlaki, Arjuna, Asoka bark, Asparagas, Babchi, Black Cumin, Calotropis, Capsicum, Cassia fruit, Castor, Cathranthus, Chaulmoogra, Cinnamon, Colocynth, Colophony, Coriander, Eucalyptus, Fenugreek, Garlic allium, Ginger, Henna, Herpestis, Hydrocotyle, Indian Ipecac, Indian Sarsaparilla, Indian Senega, Indian Squill, Kalamegh, Kurchi bark, Lemongrass, Linseed, Myrobalan, Neem, Nux-vomica, Papaya, Peppermint, Rauwolfia, Sesame, Stramonium, Turmeric, Vasaka, and Withania (Ghani, 1998).

Table 1.1: Examples of Crude Drugs and Their Therapeutic Uses (Ghani, 1998)

Drugs	Plant Source	Therapeutic Use
Digitoxin, digoxin	*Digitalis purpura, Digitalis lanata*	Cardiotonic
Morphine	*Papaver somniferum*	Sedative, narcotic, analgesic
Quinine, quinidine	*Cinchona sp.*	Antipyretic, antimalarial
Vinblastin, vincristine	*Catharanthus roseus*	Anticancer
Paclitaxel	*Texus brevifolia*	Anticancer
Theophylline	*Camellia sinensis*	Smooth muscle relaxant
Reserpine, Rescinnamine	*Rauwolfia sp.*	Hypotensive, vasodilator
Menthol	*Mentha piperita*	Anti-pruritic, antiseptic
Colchicine	*Colchicum autumnale*	Anti-gout, anti-arthritic
Hyoscine, Hyoscyamine	*Datura, Hyoscyamus, Scopolia, Duboisia spp.*	Parasympatholytic, Mydriatic
Papaverine	*Papaver somniferum*	Smooth muscle relaxant
Pilocarpine	*Pilocarpus jaborandi*	Parasymparhomimetic, cholinergic
Theobromine	*Theobroma cacao*	Smooth muscle relaxant,

1.8 Rationale of the Work

The use of medicinal herb in the treatment and prevention of disease is attracting attention by scientists' worldwide (Sofowora, 1982). This is corroborated by World Health Organization in its quest to bring primary health care to the people. The plant kingdom has long serve as a prolific source of useful drugs, foods, additives, flavoring agents, colorants, binders, and lubricants. As a matter of fact, it has been estimated that about 25% of all prescribed medicine today are substances derived from plants. Chemicals that make a plant valuable as medicinal plant are:

1. Alkaloids (Compound has addictive or pain killing or poisonous effect and sometimes help in important cures),
2. Glycosides (Use as heart stimulant or drastic purgative or better sexual health),

3. Tannins (Used for gastrointestinal problems like diarrhea, dysentery, ulcer, wounds, and for skin diseases),
4. Volatile/ Essential oils (Enhance appetite and facilitate digestion or use as antiseptic/ insecticide and insect repellant properties),
5. Fixed oils (present in seeds and foods can diminish gastric or acidity)
6. Gum-resins and mucilage (possess analgesic property that suppress inflammation and protect affected tissues against further injury and cause mild purgative) and
7. Vitamins and minerals (Fruits and vegetable are the sources of vitamins and minerals, and these are used popularly in herbals (www.*Life.umd.edu*).

According to WHO, medical plants are accessible, affordable and culturally appropriate source of primary health care for more than 80% of Asia's population (Hossain M.Z.). Bangladesh is an Asian country where only 20% of the people can be provided with modern healthcare services while the rest 80% are dependent on traditional plant-based systems. The use of traditional medicine is increasing in developing countries. This is probably due to the escalating in population, to the government supports to the forms of indigenous medicine, and finally, to the patriotic desire to revive and maintain the traditional culture. There are several studies on the botanical aspects of the plants of Bangladesh. No accurate result has been published regarding the number if medicinal plants in Bangladesh. Due to favorable climate, abundant rainfall and fertile soil, plants are sufficient in our country. Almost 5000 plants spp found in Bangladesh, about 1000 spp are said to have medicinal qualities. Recent study has identified about 550 medicinal plants in Bangladesh (Yusuf, 1994).The chemical ingredients and uses of 449 medicinal plants have been enlisted (Ghani, 1998). According to concerned authorities, much more medicinal plant species are still waiting to be enlisted as the important medicinal plants of Bangladesh. Day by day phytochemical studies of medicinal plants have got a rapid place with technological advancement. Many chemical compounds of diversified nature from plants often played an important role to give a new direction for laboratory synthesis of many new classes of drug molecules. Literature review reveals that limited phytochemicals and biological investigations have been carried out on the rhizomes of *Hedychium coronarium*.

1.9 Aim of the Present Work

Hedychium coronarium (Bengali name Dolan Chapa, Family: Zingiberaceae) is a perennial plant and widely cultivated in Japan, India, South China and South Asian countries including Bangladesh. Its rhizomes have been used for the treatment of headache, contusion inflammation and sharp pain due to rheumatism. Various cytotoxic diterpenes, farnesanetype sesquiterpenes and labdane-type diterpenes were isolated and characterized from the rhizomes of *Hedychium coronarium*. Anti-inflammatory, analgesic, antihypertensive, diuretic, leishmanicidal and antimalarial activities of rhizomes of this plant, have also been reported by several researchers. In present study, we report the analgesic, central nervous system depressant and cytotoxic effects of methanol extracts prepared successively from rhizomes of *Hedychium coronarium*.

1.10 Plan of the Present Work

The following aspects of *Hedychium coronarium* (rhizomes) have been intended to be accomplished in this research work:

- Successive extraction of dried rhizomes with methanol.
- Phytochemical Screening
- Investigations for analgesic activity
- Screening for central nervous system (CNS) depressant activity
- Brine shrimp lethality bioassay

CHAPTER 2: LITERATURE REVIEW

2.1 Plant Review

Common names of *Hedychium coronarium*

Butterfly Ginger Lily, White Ginger Lily, Graland Flower, Dolan Champa (Beng.)

Botanical Classification of *Hedychium coronarium*

Kingdom: Plantae
  Subkingdom: *Viridaeplantae*
    Division: Magnoliophyta
      Class: *Liliopsida*
        Subclass: *Commelinidae*
          Order: *Zingiberales*
            Family: Zingiberaceae
              Genus: *Hedychium*
                Species: *H. coronarium*

Description

White butterfly ginger lily, or simply ginger lily, is a tropical perennial and a cousin of culinary ginger (*Zingiber officinale*) of bread, snap and ale fame. Its green stalks grow from thick rhizomes to a height of 3-7 ft (0.9-2.1 m). Leaves are lance-shaped and sharp-pointed, 8-24 in (20-61 cm) long and 2-5 in (5-12.7 cm) wide and arranged in 2 neat ranks that run the length of the stem. From midsummer through autumn the stalks are topped with 6-12 in (15.2-30.5 cm) long clusters of wonderfully fragrant white flowers that look like butterflies. The flowers eventually give way to showy seed pods chock full of bright red seed

Geographic Distribution

Hedychium coronarium (Zingiberaceae), which has many common names such as butterfly ginger, butterfly lily, cinnamon jasmine, garland flower, and ginger lily.*Hedychium coronarium*, is a native to tropical and semi-tropical areas of Asia and Himalayas. Ginger lily is considered an invasive weed in Brazil. It was brought to the country by African slaves during the slavery era there. It is also known as the National Flower of Cuba 'Flor de Mariposa' or butterfly flower because it looks like a flying white butterfly. It is also widely cultivated in India, South China, Japan, and Southeast Asian countries including Bangladesh (www.google.com)

Chemical constituents and properties

Dried rhizome contains: starch, 3 %; glucose, 4.5 %; albumen, 1.6 %; fats, 0.33%; resinous acid, 3.6%; resin, 5.9 %; gum, 13.7 %; organic acids, essential oil. The flower yields a fragrant essential oil; the rhizome, a volatile oil. Decoction of the rhizome is anti-rheumatic, tonic and excitant. In Ayurveda, it's considered febrifuge, tonic, stimulant and antirheumatic. (Philippine medicinal plants)

Medicinal Properties and Uses

- Skin diseases, especially with pruritus: Reduce the leaves to a paste and apply tolerably warm to areas of affected skin.
- Post-partum and rectal inflammation: Infusion of leaves.
- Mumps, acne, and localized rheumatic complaints: Paste of leaves applied to affected areas.
- Warm paste of leaves also used for pruritus.
- Cough and thrush: Infusion of flowers, 40 grams to a pint of boiling water, 4 glasses of tea daily.
- Fever: Fruit as a cooling drink.
- The fruit has been used for a variety of maladies: beriberi, cough, prevention of scurvy.
- Infusion of leaves also drank as a protective tonic after childbirth.

Other uses

Culinary: Young buds and flowers are edible. Used as flavoring. Roots used as famine food. Folkloric: Decoction of stems near the rhizome used as a gargle for tonsillitis; or the raw stem chewed for same purpose. The juice of the stem applied externally for swellings. In India, it is sold in bottles of extract called Gulbakawali Ark; used as eye tonic and for to prevent eye cataracts and in Chinese medicine, used for headache, inflammatory pains, rheumatism. In the provinces, the fragrant flowers popular in the making of wreaths and bridal bouquets. Stems are 45% cellulose, used in making paper. (Philippine medicinal plants).

Members of the genus *Hedychium*

ZipcodeZoo has pages for 244 species, subspecies, varieties, forms, and cultivars in this genus. Here are just 100 of them:

H. acuminatum · *H. album* · *H. angustifolium* · *H.* 'Anne Bishop' · *H.* 'Anne S. Bishop' (Flowering Ginger) · *H.* 'Apricot' · *H. aurantiacum* (Orange Bottlebrush Ginger) · *H. aureum* · *H.* 'Ayo' (Hardy Ginger) · *H. barbatum* · *H.* 'Beni-Oiran' (Ginger) · *H.* 'Beni-oran' · *H.* 'Betty Ho' · *H. biflorum* · *H. bijiangense* · *H. bipartitum* · *H. boloveniorum* · *H. bordelonianum* · *H. borneense* · *H. bousigonianum* · *H. breviaule* · *H. calcaratum* · *H. candidum* · *H. carneum* · *H.* 'Carnival' · *H. cernuum* · *H. chrysoleucum* · *H.* 'Cinnabar' · *H. coccineum* (Orange Bottlebrush Ginger) · *H. coccineum* Buch.-Ham. ex Sm. var. *carneum* (Roscoe) Baker · *H. coccineum* 'Disney' · *H. coccineum* 'Orange Brush' · *H. coccineum* 'Tara' (Orange Bottlebrush Ginger) · *H. coccineum* var. *angustifolium* · *H. coccineum* var. *angustifolium* 'Peach' · *H. coccineum* var. *carneum* · *H. collinum* · *H. convexum* · *H. coronarium* (Butterfly Ginger Lily) · *H. coronarium* 'Andromeda' · *H. coronarium* 'Gold Spot' · *H. coronarium* 'Orange Spot' · *H. coronarium* var. *coronarium* · *H. coronarium* var. *flavescens* · *H. coronarium* var. *maximum* · *H. coronarium* var. *urophyllum* · *H. coronarium* var. *flavum* · *H. coronarium* × *ellipticum* · *H. coronarium* × *gardnerianum* · *H. crassifolium* · *H. cylindricum* (Epiphytic Hedychium) · *H.* 'Daniel Weeks' · *H.* 'Dave Case' · *H. deceptum* · *H. dekianum* · *H. densiflorum* · *H. densiflorum* 'Assam Orange' (Hardy Ginger Lily) · *H. densiflorum* 'Sorung' · *H. densiflorum* 'Stephen' (Hardy Ginger Lily) · *H. denticulatum* · *H.* 'Devon Cream' · *H.* 'Doctor Moy' · *H.* 'Double Eagle' · *H.* 'Dr. Moy' (Hardy Ginger Lily) · *H. efilamentosum* · *H. elatum* · *H.* 'Elizabeth' (Hardy Ginger Lily) · *H. ellipticum* (Hedychium) · *H. ellipticum* red bracts · *H. elwesii* · *H. emeiense* · *H. erythrostemon* · *H.* 'F.W. Moore' · *H. fastigiatum* · *H.* 'Filigree' · *H. flavescens* (Cream Garland-Lily) · *H. flavum* · *H. flavum* 'Royale' · *H. formosum* · *H. forrestii* (Ginger Lily) · *H. forrestii* var. *latebracteatum* · *H.* 'Gahili' · *H. gandasulium* · *H. gardneranum* · *H. gardnerianum* (Kahila Garland-Lily) · *H. gardnerianum* 'Kahili Fiesta' (Kahili Ginger) · *H. gardnerianum* 'Roscoe' · *H. gardnerianum* var. *pallidum* · *H. gardnerianum* × *spicatum* · *H. gardnerianum* yellow-flowered · *H.* 'Giant Yellow' · *H. giganteum* · *H. glabrum* · *H. glaucum* · *H.* 'Golden Butterfly' · *H.* 'Golden Glow' · *H.* 'Gold Flame' · *H. gomezianum* · *H. gracile* · *H. gracillimum*

Figure 2.1: **Plant of *Hedychium coronarium*** (*Hedychium coronarium, lirio-do-brejo*)

Figure 2.2 : **Leaves of *Hedychium coronarium*** (*Hedychium coronarium,* lirio-do-brejo)

Figure 2.3: Rhizomes of *Hedychium coronarium* (Garden Plants, *Hedychium coronarium*, CSID)

Figure 2.4: White color flower of *Hedychium coronarium* (Godofredo and Stuart, 2016)

Figure 2.5: Whole part of *Hedychium coronarium* (Godofredo and Stuart, 2016)
Literature Review
2.2 Phytochemical properties

Hedychium coronarium contain many bioactive compounds including saponins, glycosides, fats and volatile oil. The main chemical found in the plant include hedychicoronarin, peroxycoronarin D,7β hydroxyl-calcaratarin A and E, 7β hydroxyl-6-oxo-labda-8, 12-diene-15, 16-dial have been isolated from the rhizomes of Hedychium coronarium (Lu et al., 2009; Verma et al; 2012). Hedychicoronarin, peroxycoronarin D were isolated as optically active colorless oil. The phytochemical study of the rhizomes from Hedychium coronarium showed the presence of benzoyl eugenol along with the labdane diterpenes isocoronarium D and ethoxy coronarin D also. Hedychium coronarium afforded oils whose major constituents were β-pinene (20%), Linalool (15.8%), α-pinene (10.1%), 1,8-cineole (10.7%) and α-terpineol (8.6%) in the leaf while the root consists mainly of β-pinene (23.6%), α-humulene (17.1%), β-caryophyllene (13.0%), α-pinene (6.9%) (Chen et al., 2013; Sing et al; 2013). The volatile constituents of the various parts of *H. coronarium* from other parts of the world have been reported (Mishra, 2013). Although ubiquitous monoterpenes and sesquiterpeens were the main components of these oils, the identities of these compounds differed from one another. This led to the delineation of various chemotypic forms of the essential oils of *Hedychium coronarium*. The compositional pattern of the leaf oil (β-pinene, linalool, α-pinene, 1,8-cineole) and the root (β-pinene, β-caryophyllene, α-humulene) in this study seems to be new chemotypic forms of essential oil of the plant when compared with previous studies (Verma et al., 2013).

Phytoconstituents

Important phytoconsttuents isolated from *Hedychium coronarium* (Lu et al., 2009; Verma et al; 2012; Chen et al., 2013; Sing et al; 2013; Verma et al., 2013; Joy et al., 2007; Nakamura et al.,2008)

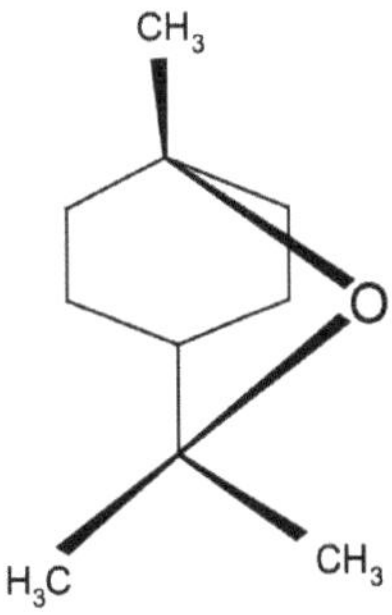

1, 8 -cineole 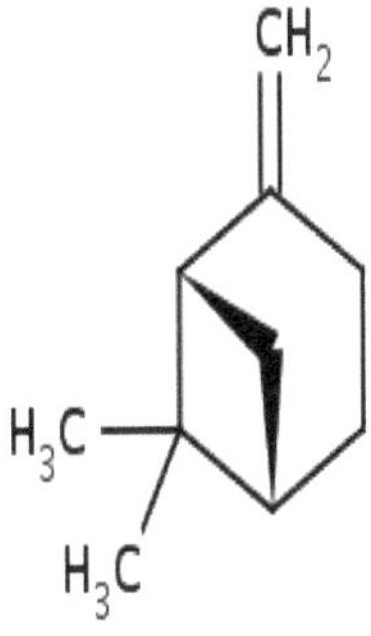   β- pinene 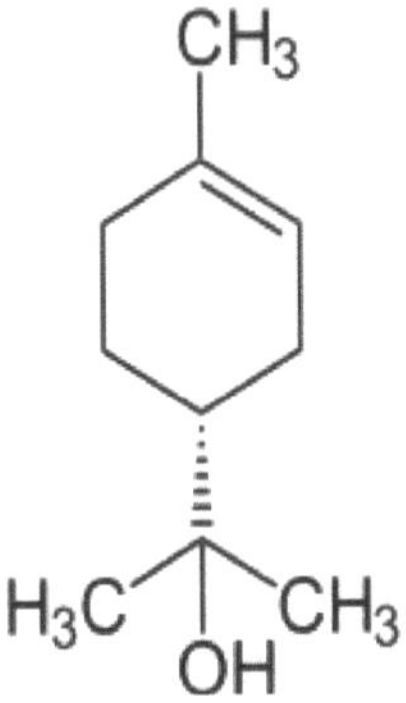   α-terpineol

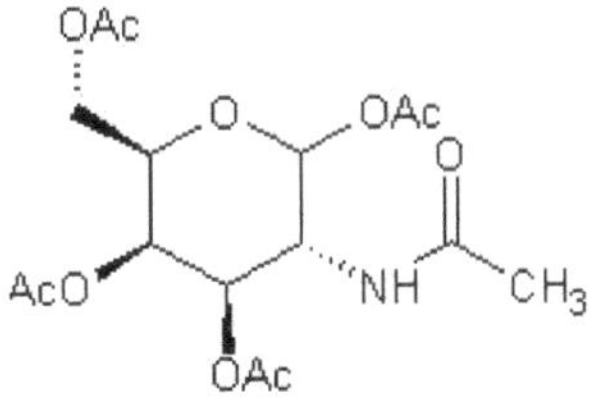

D-galactosamine

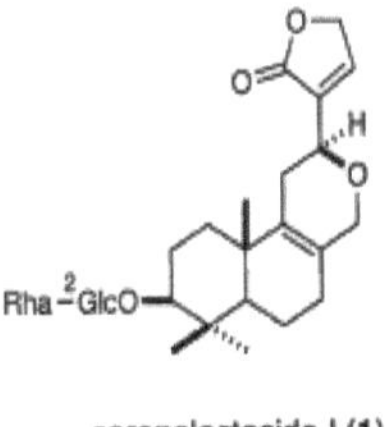

coronalactoside I (1) 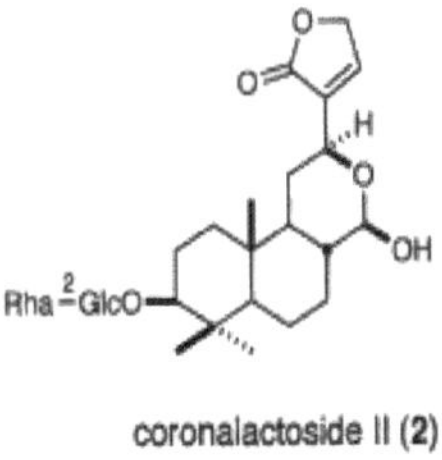   coronalactoside II (2) 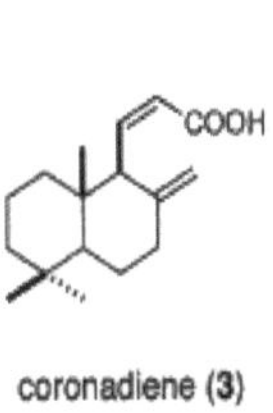   coronadiene (3)

Figure 2.6: structure of chemical compounds isolated from *Hedychium coronarium*

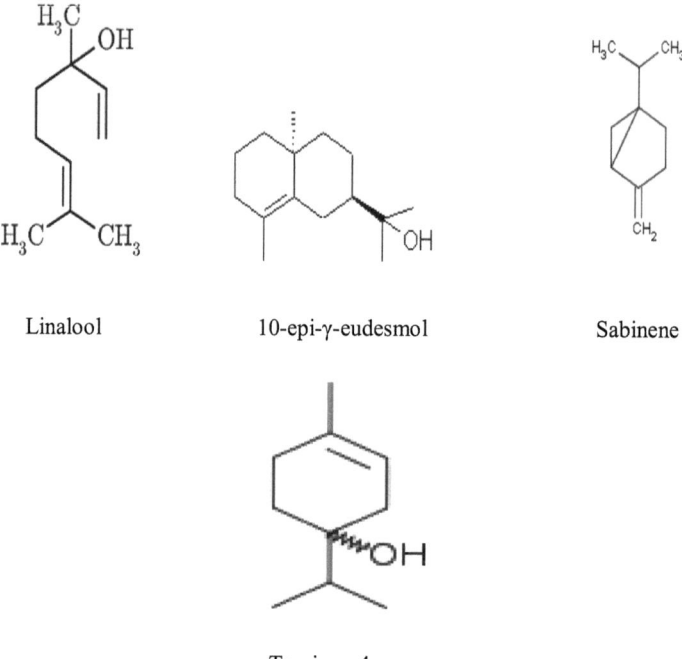

Figure 2.6: structure of chemical compounds isolated from *Hedychium coronarium* (Lu et al., 2009; Verma et al; 2012; Chen et al., 2013; Sing et al; 2013; Verma et al., 2013; Joy et al., 2007; Nakamura et al.,2008)

2.3 Pharmacological properties

Its rhizome is used in the treatment of diabetes. Traditionally it is used for the treatment of tonsillitis, infected nostrils, tumor and fever. The rhizomes of this plant have antimicrobial activity (Joy *et al.*, 2007), anti-inflammatory and analgesic activity (Shrotriya *et al.*, 2007; Dash et al., 2011), CNS depressant activity (Dixit *et al.*, 1979; Dash et al., 2011), antibacterial and cytotoxic activity (Aziz *et al.*, 2009; Dash et al., 2015) and inhibitory activity (Morikawa *et al.*, 2002).The flowers extract of this plant have anti-inflammatory and antioxidant activity (Lu *et al.*, 2009) and induced cytotoxicity (Nakamura *et al.*, 2008) .It is also used as anti-rheumatic, antioxidant, excitant, febrifuge, and tonic (Verma et al., 2013). It has been reported that the essential oil extracted from leaves, flowers and rhizome of the plant have molluscicidal activity, potent inhibitory action, antimicrobial activities, antifungal, anti-inflammatory, anti-bacterial and analgesic effects. The seeds are aromatic, carminative and stomachic. The plant also possessed analgesic and neuropharmacological, anti-inflammatory, antimicrobial and cytotoxic activities (Ramarao et al., 1990). This plant has tremendous medicinal properties and its various parts are used in traditional as well as modern medicine. The rhizome of the plant is used in the treatment of diabetes, cold, body aches, headache, lancinating pain, contusion, inflammation and rheumatic pain. The rhizome has anti-cancerous, anti-hypertensive, diuretic, leishmancidal, anti-malarial activities and also in irregular menstruation, piles bleeding and stone in urinary blood (Thanh et al., 2014).

CHAPTER 3: MATERIALS AND METHODS

All materials and methods have been adopted in the Pharmacology Laboratory of the Department of Pharmacy, Stamford University Bangladesh.

3.1 Plant material
Hedychium coronarium was selected for investigation. The rhizome was used for research purpose.

Collection of the Plant
The Rhizomes of *Hedychium coronarium* was collected from the local area of Mirpur, Dhaka during the fast weak of January 2010. Dust, dirt and the undesirable materials were then separated manually.

Identification
The collected plant was then identified by Bangladesh National Herbarium, Mirpur, Dhaka. Date of investigation by the herbarium was January 4, 2010. Accession code of the plant given in the following table for further reference.

Table 3.1: Accession Code of the Plant

Local Name	Botanical Name and Family	Accession Code
Dolan Champa	*Hedychium coronarium* (Zingiberaceae)	34,484

Preparation of the plant sample
The collected and identified plant's rhizome was cut into small pieces separately and then dried in the sun and finally dried in a hot air oven at 35-40° C for 24 hours. After complete drying the rhizome was reduced to coarse powder separately with the help of a mechanical grinder and the powder was stored in a suitable container for extraction. The dried grinded powder weighed by rough balance.

Figure 3.1 Grinding by Using a Blender

Extraction of Rhizomes

The plant parts were extracted by cold extraction method. 200 gm powder was obtained from the plant rhizomes. A glass made jar with plastic cover was taken and washed. The jar was rinsed with methanol and dried. Then the dried powder 150gm was taken into the jar. After that 450 ml of 80% of methanol was poured into the jar up to 1 inch height above the sample surface as it can sufficiently cover the sample surface. The plastic cover with aluminum foil was closed properly to resist the entrance of air into the jar. This process was performed for 5 days. The jar was shaken in several times during the process to get better extraction. The extract was separated from the plant debris by filtration by filter paper. The extract was concentrated by evaporation and dried to solid in an oven.

Figure 3.2 : Rhizomes extract of *Hedychium coronarium*

Overall Extraction Process

The Flow Chart of the extraction process is given below-

Collection of plants at suitable time and session
↓
Identification of plants/plants part
↓
Drying of rhizome in a suitable size
↓
Grinding
↓
Powders were collected and stored in a cool and dry place
↓
Preparation of extracts with methanol by maceration
↓
Evaporation and drying of methanolic extract
↓
Methanol rhizome extract of *Hedychium coronarium*

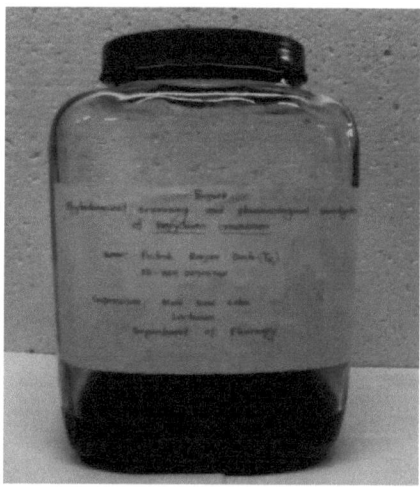

Figure 3.3 : Extraction of *Hedychium coronarium* by maceration

3.2 Materials Used in the Study

For the accomplishment of this study several materials are used. The materials which are used in this study are arranged in their category

List of Apparatus Used

Different types of apparatus are used for the extraction and experimentation which are listed below-

Table 3.2: List of Apparatus Used for the Experiment

Serial No.	Apparatus Name	Source
1	Vortex Mixture	China
2	LC Oven	LAB-LINE USA
3	Syringe	Opsosaline Ltd. Bangladesh
4	Feeding Needle	Local made
5	Digital Weighing Balance	Denver Instrument Company USA
6	Cotton	India
7	Blender	NOWAKE, Japan
8	Aluminium Foil	DIAMOND, USA
9	Filter Paper	11cm DIA, China
10	Marker Pen	RED LEAF, Japan
11	Thermostatic Water Bath	Shanghai, China

List of Glassware Used

Different types of glassware are used during experimentation, major wares are listed below-

Table 3.3: List of Glassware Used

Serial No.	Apparatus Name	Source
1	Volumetric Flask (10ml & 50ml)	Changdu, China
2	Test tube	Changdu, China
3	Beaker	Changdu, China
4	Funnel	Changdu, China
5	Measuring Cylinder	Changdu, China
6	Glass Rod	Changdu, China

List of Reagents Used

Many chemicals are used as solvent which are listed in Table-3.4.

Table 3.4: List of Reagents Used for the Experiment

Serial No.	Name of Reagents	Source
1	Methanol	Merck, Germany
3	Acetic Acid	Merck, Germany
4	Tween 80	India
5	Distilled Water	Laboratory Prepared
6	Diclofenac-Na	India
7	Castor oil	Spain
8	Dimethyl sulfoxide (DMSO)	India

Animal Used

A special type of mice has been used for the experiment. Details of the mice are given in the following table.

Table 3.5: Detail information of the mice used for the Experiment

Serial No.	Name of the Animal	Source
1	*Swiss albino* mice Average weight: 25 gm.	ICDDR, B, Animal House, Mohakhali, Dhaka, Bangladesh

Figure 3.4: *Swiss albino* **mouse**

Animal Feed Used

The mice were given special type of chocolate food which was supplied by the ICDDR, B.

Table 3.6: Type of Food Used for the Mice

Serial No.	Name of the Animal Feed	Source
1	Mice Pellets (Chocolate Food)	ICDDR, B, Animal House, Mohakhali, Dhaka, Bangladesh

Table 3.7: Materials Used for Animal House

Serial No.	Apparatus Name	Source
1	Polyvinyl Cages	ICDDR, B, Animal House, Mohakhali, Dhaka, Bangladesh
2	Soft Wood for Bedding of Animals	ICDDR, B, Animal House, Mohakhali, Dhaka, Bangladesh

Identification of experimental animals

Each group consists of five mice. As it was difficult to identify and observe at a time five mice receiving same treatment. Thus it was important to identify individual animal of a group during the treatment. To denote individual animal, they were marked or coded as I, II, III, IV and V.

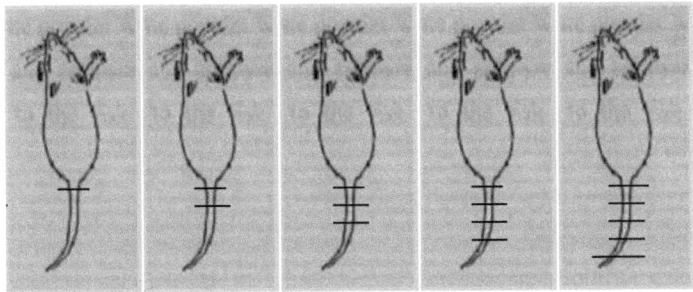

Figure 3.5. Identification of experimental animals.

3.3 PHYTOCHEMICAL SCREENING

Phytochemical Screening

The phytochemical investigation or screening is a process for the detection and evaluation of plant constituents through chemical analysis; phytochemical screening is co-related with phytochemical study. The compounds isolated through phytochemical study are applied on treated animal to find out the pharmacological effect either beneficial or toxic and thus toxic plants are separated. (Journal of Pharmaceutical Science, 1966).

In this research work methanol and aqueous extracts of *Hedychium coronarium* were screened for carbohydrates, glycosides, saponins, flovanoids, tannins, steroids, alkaloids these have pronounced medical value.

Preparation of the extract

The crude extracts were made free from pigment (decolorized) and other impurities by filtering through activated charcoal.

Preparation of Sample Solution

Small amounts of dried, decolorized extracts were appropriately treated to prepare sample solution and then subjected to various phytochemical tests.

Chemical Group Tests

Testing of different chemical groups present in Rhizome extract of *Hedychium coronarium* represent the preliminary phytochemical studies. Reagents used for different group tests are listed in the following table:

Table 3.8: Reagents used for different group tests

Chemical groups	Reagents	Tests
Carbohydrates	Molisch Reagent	Molisch's test
	Fehling's Reagent	Fehling's test
Tannins	10% Potassium dichromate	Potassium dichromate test
	5% Ferric chloride	Ferric chloride test
	1% Lead acetate	Lead acetate test
Flavonoids	Zinc ribbon and conc. hydrochloric acid	Test for flavonoids

Saponins	Water	Frothing test
Protein	Millons reagent 10% sodium hydroxide and 3% copper sulphate	Millons test Biuret's test
Steroids	Sulphuric acid	Salkowski test
Alkaloids	Mayer's reagent Dragendroff's reagent Hager's reagent	Mayer's test Dragendroff's test Hager's test
Glycosides	Aqueous sodium hydroxide	General test for glycosides
Glucosides	Fehling's reagent and Sulphuric acid	Test for Glucosides

Preparation of reagents used for different chemical group tests

Reagents were prepared following standard procedure as described by Ghani A, 2005.

Mayer's Reagent: 1.36 gm. mercuric iodide in 60 ml of water mixed with a solution contains 5 gm. of potassium iodide in 20 ml of water. Then volume is adjusted to 100 ml.

Dragendroff's Reagent: 1.7 gm. basic bismuth nitrate and 20 gm. tartaric acid were dissolved in 80 ml water. This solution was mixed with a solution contains 16 gm. potassium iodide and 40 ml water and diluted 10 times with 10% picric acid before use.

Hager's Reagent: A 1% solution of picric acid in water.

Fehling's solution A: 34.64 gm. copper sulphate was dissolved in a mixture of 0.50 ml of sulfuric acid and sufficient water to produce 500 ml.

Fehling's solution B: 176 gm. of sodium potassium tartarate and 77 gm. of sodium hydroxide in sufficient water to produce 500 ml. Equal volume of above solution was at the time of use.

Benedict's reagent: 1.73 gm. cupric sulphate, 1.73 gm. sodium citrate and 10 gm. anhydrous sodium carbonate are dissolved in water and the volume made up to 100 ml with water.

Molisch Reagent: 2.5 gm. of pure α-Naphthol dissolved in 25 ml of ethanol.

Test procedure for identifying different chemical groups

Chemical groups were identified by characteristic color changes using standard procedure (Ghani, 2003)

Tests for Carbohydrates

- **Molisch's test (General test for Carbohydrates)**

2ml solution of the extract of the plant material was taken in a test tube. 2 drops of freshly prepared 10% alcoholic solution of α-Naphthol was taken in test tube and thoroughly mixed. 2ml of conc. Sulphuric acid was given to flow down the side of the inclined test tube so that the acid forms a layer beneath the aqueous solution. A reddish violet ring was formed at the junction of the two layers if a carbohydrate was present. A dark purple solution was formed on standing or shaking.

The test tube was shaken and allowed to stand for 2 minutes, and then it was diluted with 5ml of water. A dull violet precipitate was formed immediately which confirmed the presence of carbohydrates.

- **Fehling's test (Standard test for Reducing sugars)**

2ml of an aqueous extract of the plant material add 1ml of a mixture of equal volumes of Fehling's solutions A and B. Boil for a few minutes. A brick-red precipitate was formed immediately which confirmed the presence of carbohydrates.

Tests for Tannins

- **Ferric chloride test**

5ml solution of the extract was taken in a test tube. Then 1ml of 5% Ferric chloride solution was added. Greenish black precipitate was formed which confirmed the presence of tannins.

- **Potassium dichromate test**

5ml solution of the extract was taken in a test tube. Then 1ml of 10% Potassium dichromate solution was added. A yellow precipitate was formed in the presence of tannins.

- **Lead acetate test**

5ml of an aqueous extract of the plant material was taken in a test tube and added a few drops of a 1% solution of lead acetate. A yellow precipitate was formed which confirmed the presence of tannins.

Test for Flavonoids

0.5 ml of an alcoholic extract of the sample was taken in a test tube. Then a small piece of zinc ribbon and 5-10 drops of conc. hydrochloric acid was added. Boil the solution for a few minutes. Development of orange to red (flavones) colour which indicates the presences of flavonoids.

Tests for Saponins

- **Frothing test**

0.5ml of alcoholic extract was diluted to 10ml with distilled water and shaken in a graduated cylinder for 3-5 minutes. Production of persistent frothing indicates the presence of saponins.

5.10 Tests for Proteins

- **Millons test**

A small amount of an aqueous extract of the plant material was taken in a test tube. Then 5-6 drops of Millons reagent was added. Formation of a white precipitate turning red on heating indicates the presences of proteins in the sample.

- **Biuret's test**

1ml of aqueous extract of the plant material was taken in a test tube. Then 5-6 drops of 10% sodium hydroxide solution and 1-2 drops of 3% copper sulphate solution were added. A red color indicated the presence of proteins.

Tests for Steroids

- **Salkowski test**

2ml solution of chloroform extract was taken and then 1ml Sulphuric acid was added. Presence of red color indicated the presence of steroids.

5.12 Tests for Alkaloids

- **Mayer's test**

2ml solution of the extract and 0.2ml of dilute Hydrochloric acid were taken in a test tube. Then 1ml of Mayer's reagent was added. Creamy white precipitate was formed and that indicated the presence of alkaloids.

- **Dragendroff's test**

2ml solution of the extract and 0.2ml of dilute Hydrochloric acid were taken in a test tube. Then 1ml of Dragendroff's reagent was added. Orange red precipitate was formed which indicated the presence of alkaloids.

- **Hager's test**

2ml solution of the extract and 0.2ml of dilute Hydrochloric acid were taken in a test tube. Then 1ml of Hager's reagent was added. Yellow crystalline precipitate was formed which indicated the presence of alkaloids.

Tests for Glycosides

- **General test for glycosides**

A small amount of alcoholic extract was dissolved in1 ml of water and adding a few drops of aqueous sodium hydroxide solution. A yellow color develops in the presence of glycosides.

Tests for Glucosides

Dissolve a small amount of an alcoholic extract of the plant material in water and alcohol and boil with Fehling's solutions A and B. A yellow color develops in the presence of glucosides. Dissolve another portion of the extract in water and alcohol, boil with a few drops of dilute sulphuric acid, neutralize with sodium hydroxide solution and boil with Fehling's solution A and B. A brick-red precipitate was formed which indicated the presence of glucosides.

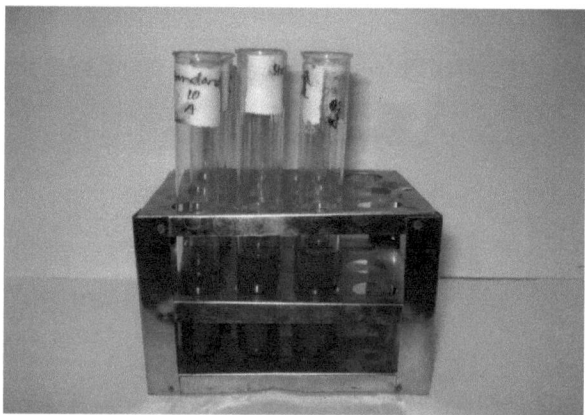

Figure 3.6: Stock solution used for chemical group tests

Results and Discussion

Results of the different chemical group tests are given in the following table

Table 3.9: Results of phytochemical screening (Dash et al., 2015)

Extract	Chemical Group	Result
Me-OH extract of *Hedychium coronarium*	Carbohydrates	+
	Tannins	
	Ferric chloride test	−
	Potassium dichromate test	−
	Lead acetate test	−
	Flavonoids	+
	Saponins	+
	Protein	−
	Steroids	+
	Alkaloids	
	Mayer's test	+
	Dragendroff's test	+
	Hager's test	+
	Glycosides	−
	Glucosides	−

Me-OH = Methanol; (+) =Presence; (−) =Absence

The results of phytochemical screening showed that the methanolic extract of the rhizomes of *Hedychium coronarium* contained carbohydrates, flavonoids, saponins, steroids, alkaloids. Alkaloids has addictive or pain killing or poisonous effect and sometimes help in important cure. Saponins may help to prevent colon cancer. Flavonoids possess antiallergic, anti-inflammatory, antiviral and antioxidant activities. Steroids are used to suppress various allergic, inflammatory and autoimmune disorders.

3. 4 PHARMACOLOGICAL INVESTIGATIONS

3.4.1 Analgesic activity test for the crude extract of *Hedychium coronarium*

Study of analgesic activity of crude extract of *Hedychium coronarium* by acetic acid induced writhing method has been adopted in the Pharmacology Laboratory of the Department of Pharmacy, Stamford University Bangladesh on mice model.

Analgesic drugs which are currently in use are either narcotics or non-narcotics which have proven side and toxic effects. To develop new synthetic compounds in this category is an expensive venture and again may have problems of side effects. On the contrary, many medicines of plant origin had been used and are in use successfully since long time without any serious effects (Ikram, 1983).

Pain has been officially defined as an unpleasant sensory and emotional experience associated with actual or potential tissue damage. Pain acts as a warning signal against disturbances of the body and has a proactive function (Tripathi, 1999).

Analgesic means a drug that selectively relieves pain by acting in the CNS or on peripheral pain mechanisms, without significantly altering consciousness. So, analgesic activity means capacity of a substance to neutralize the pain sensation (Rang, 1993).

The lack of potent analgesic and anti-inflammatory drugs now actually in use prompted the present study, in which rhizomes of *Hedychium coronarium* had been selected for its reported biological activities in indigenous system of medicine.

Acetic acid induced writhing test

Objectives
The purpose of this study was to examine analgesic effect of methanolic extract of rhizome of *Hedychium coronarium* on mice in a peripheral model of analgesic activity test.

Principle
The acetic acid induced writhing method is an analgesic behavioral observation assessment method that demonstrates a noxious stimulation in mice. The test consists of injecting the 0.7% acetic acid solution intraperitoneally and then observing the animal for specific contraction of body referred as writhing. A comparison of writhing was made between

positive control (diclofenac), control and test sample given orally 30 minutes prior to acetic acid injection. If the sample possesses analgesic activity, the animal that received the sample will give lower number of writhing than the control, i.e. the sample having analgesic activity will inhibit writhing. Diclofenac-Na is used as standard drug.

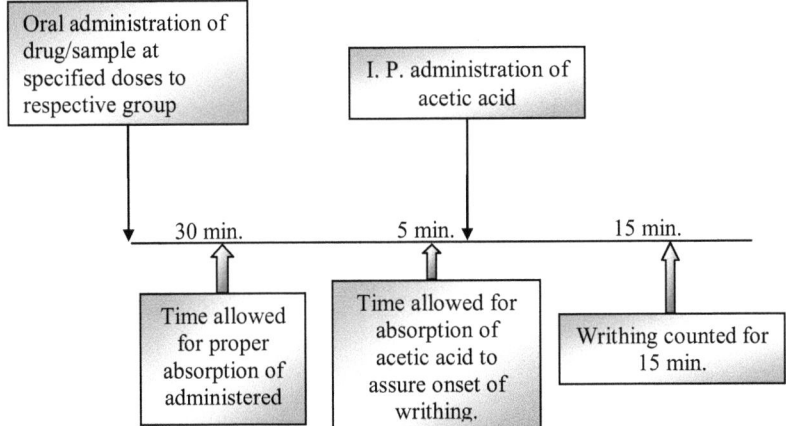

Figure 3.7: Schematic representation of acetic acid induced writhing of mice for investigation of analgesic activity (Dash et al., 2013)

Experimental Animal

Young *Swiss albino* mice aged 3-4 weeks, average weight 25 gm. were used for the experiment. The mice were purchased from the animal research branch of the International Centre for Diarrhoeal Diseases and Research, Bangladesh (ICDDR, B). For this experiment, five groups (I, II, III, IV and V) of mice were used and each group contains 5 mice.

Experimental Design

The method described by Howlader *et al.* (2006) was adopted to study the effect of the *Hedychium coronarium* extract on acetic acid induced writing test. Test samples and control were given orally by means of a feeding needle. A thirty (30) minutes interval was given to ensure proper absorption of the administered substances. Then the writhing inducing chemical, acetic acid solution (0.7%) was administered intraperitoneally to each of the animals of a group. After an interval of fifteen (15) minutes, this was given for absorption and no of squirms (writhing) was counted for 5 minutes.

Preparation of the Test Materials

To prepare solution of the plant extract at a doses of 100 mg/kg, 200mg/kg and 400mg/kg body weight, 180 mg extract was measured and added with it 9 ml of distilled water and mixing with the help of vortex apparatus. From this solution 5ml/kg was taken for 100mg/kg, 200mg/kg and for 400 mg/kg dose respectively.

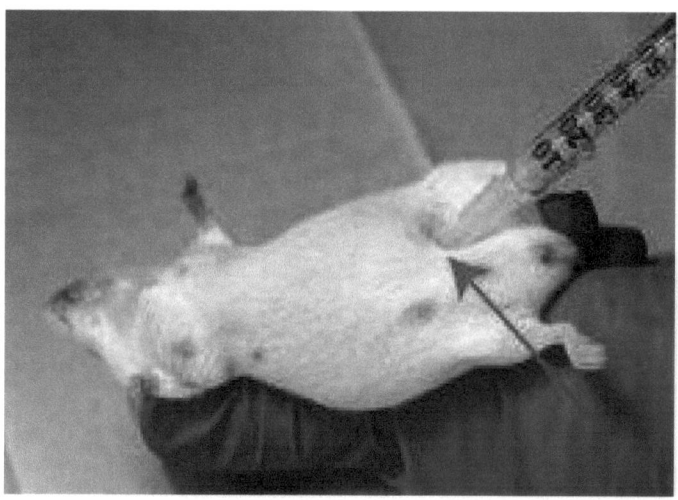

Figure 3.8: Preparing mice for writhing by injecting acetic acid (Dash et al., 2013)

Mechanism of pain induction in acetic acid induced writhing method

Acetic acid is a pain stimulus. Intraperitoneal administration of acetic acid (0.7%) causes localized inflammation. Such pain stimulus causes the release of free arachidonic acid from tissue phospholipids by the action of phospholipase A_2 and other acyl hydrolases. There are three major pathways in the synthesis of the eicosanoids from arachidonic acid. All the eicosanoids with ring structures, which is the prostaglandins, thromboxanes and prostacyclines, are synthesized via the cyclooxygenase pathway. The released prostaglandins, mainly prostacyclines (PGI_2) and prostaglandin-E have been reported to be responsible for pain sensation by exciting the Aδ-fibers. Activity in the A δ-fibers cause a sensation of sharp well localized pain (Rang *et al*, 1993). Diclofenac used as the standard in this method, acts by inhibition of prostaglandin synthesis. Any agent that lowers the number of writhing will demonstrate analgesia by inhibition of prostaglandin synthesis.

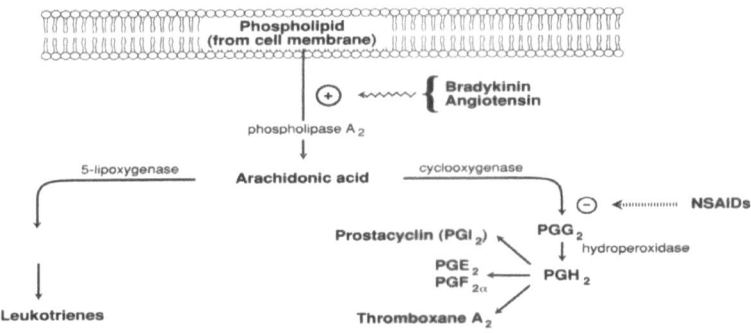

Figure 3.9: Synthesis of Prostaglandins and Leukotrienes (Dash et al., 2013)

Study Design

Experimental animals were randomly selected and divided into five groups denoted as group-I, group-II, group-III, group-IV and group-V consisting of 5 mice in each group. Each group received a particular treatment i.e. control, standard and the three doses of the extract. Each mouse was weighed properly and the doses of the test samples and control materials were adjusted accordingly.

Table 3.10: Experiment Profile to assess the effect of crude extract of *Hedychium coronarium* on acetic acid induced writhing of mice

Animal Group	Treatment	No. of Animals	Dose	Route of Administration
Control	Water	5	5ml/kg	Oral
Standard	Diclofenac-Na	5	25 mg/kg	i.p
Group-I	Methanolic rhizome extract of *Hedychium coronarium*	5	100 mg/kg	Oral
Group-II	Methanolic rhizome extract of *Hedychium coronarium*	5	200 mg/kg	Oral
Group-III	Methanolic rhizome extract of *Hedychium coronarium*	5	400 mg/kg	Oral

Counting of Writhing

Each mouse of all groups were observed individually for counting the number of writhing they made in 15 minutes commencing just 5 minutes after the intrperitoneal administration of acetic acid solution. Full writhing was not always accomplished by the animal, because sometimes the animals started to give writhing but they did not complete it.

Figure 3.10: Half Writhing Given by Mice

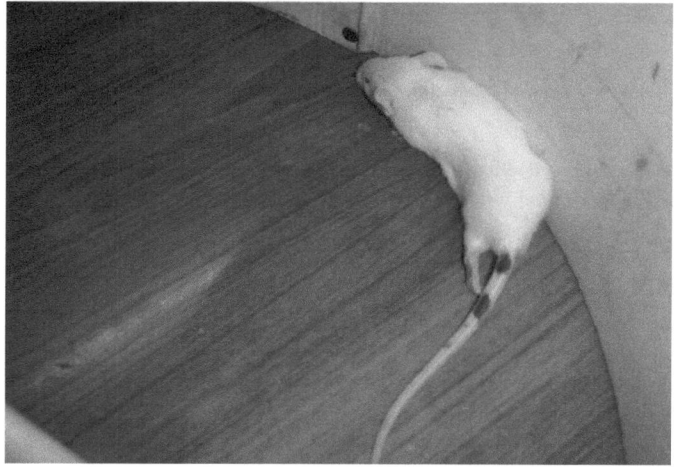

Figure 3.11: Full Writhing Given by Mice

Results

Table 3.11: Effect of methanol extract of *Hedychium coronarium* on acetic acid induced writhing test in mice. (Dash et al., 2011)

Group	Treatment and Dose	Writhings (Mean ± SEM)	% of writhing	% of writhing inhibition
Control	Water (5 ml/kg)	41.3±1.32	100.00	0
Standard	Diclofenac sodium (25 mg/kg i.p.)	10.0±0.42**	24.21	75.78
Group-I	Extract (100 mg/kg per oral)	30.5±1.035**	73.85	26.15
Group-II	Extract (200 mg/kg per oral)	21.5±0.995**	52.06	47.94
Group-III	Extract (400 mg/kg per oral)	11.1±2.88**	26.87	73.12

Diclofenac sodium was administered 30 min before 0.7% acetic acid administration. Writhing was counted for 15 min, starting 5 min after acetic acid administration; **$P<0.001$ vs. control, values are mean ±SEM ($N=5$).

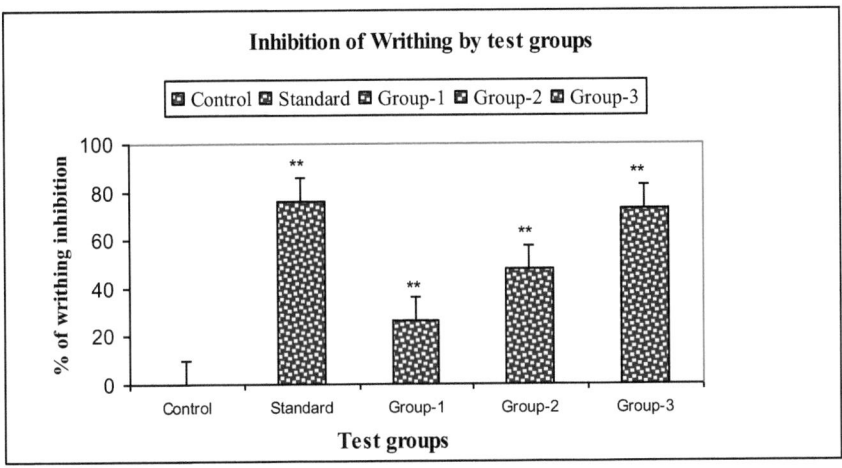

Diclofenac sodium was administered 30 min before 0.7% acetic acid administration. Writhing was counted for 15 min, starting 5 min after acetic acid administration; **$P<0.001$ vs. control, values are mean ±SEM ($N=5$

Figure 3.12: Percentage inhibition of writhing reflex by *Hedychium coronarium* (Dash et al., 2011)

Tail Immersion Test

There are two variants of the tail immersion test. One consists of applying radiant heat to a small surface of the tail. The other involves immersing the tail in water at a predetermined temperature. Although apparently similar, these two alternatives are actually quite different at a physical level.

Principle

Immersion of an animal's tail in hot water provokes an abrupt movement of the tail and sometimes the recoiling of the whole body. Again, it is the reaction time that is the time to flick the tail from hot water which is monitored. If a sample contains any analgesic principle it increases the ability of the mice to retain its tail in the hot water which is reflected in the increase in the tail flicking time (Toma *et al.*, 2003).

Preparation of the Test Materials

To prepare solution of the plant extract at a doses of 100mg/kg , 200mg/kg and 400mg/kg body weight, 180mg extract was measured and added with it 9 ml of distilled water and mixing with the help of vortex apparatus. From this solution 5ml/kg was taken for 100mg/kg, 200mg/kg and 400mg/kg dose respectively.

Procedure

The Tail immersion test was used with modification described by Ahmad *et al.*, (1992). The screening cut-off time was 5 sec, while the test cut-off time was 10sec. The extract was administered orally at three doses (100, 200, 400mg/kg body weight) using Diclofenac Sodium as standard. The post drug reaction times were measured at 0, 30, 60 and 90 minutes later. The tail of the mouse was immersed to a constant level (3 cm) in a water bath maintained at $55 \pm 0.5°$. The time to flick the tail from water (reaction time) was recorded. A maximum immersion time of 10 sec. was maintained to prevent thermal injury to the animals. A significant increase in reaction time compared with control animals was considered a positive analgesic response.

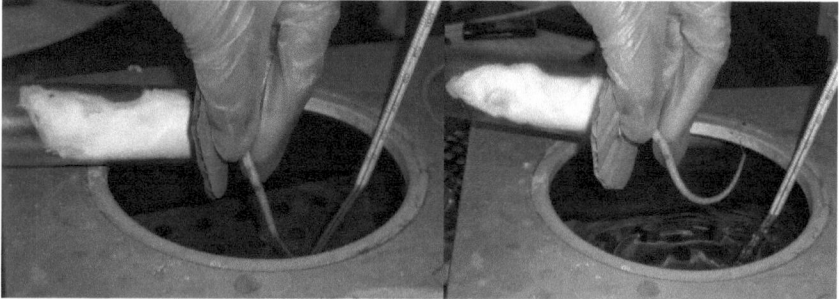

Figure 3.13: Tail immersion test on mice

Results

Table 3.12: Effects of the methanolic extract of *Hedychium coronarium* on tail immersion test. (Dash et al., 2011)

Groups	Dose (mg/kg)	Mean reaction time (s) before and after drug administration (% of tail flick elongation)			
		0 min	30 min	60 min	90 min
Control	5ml/kg	1.73±0.125	1.60±0.125	1.47±0.17	1.33±0.105
Standard	25	2.53±0.29	5.33±0.235** (69.98%)	7.39±0.07** (80.10%)	8.8±0.17** (84.88%)
Group-I	100	1.82±0.02	4.45±0.385** (64.04%)	6.09±0.405** (75.86) %	7.06±0.50** (81.16%)
Group-II	200	1.86±0.035	5.53±0.335** (71.06%)	6.28±0.495** (76.59%)	7.44±0.305** (82.12%)
Group-III	400	1.82±0.005	6.50±0.24** (75.68%)	7.38±0.325** (80.08%)	9.88±0.495** (86.53%)

Control: animals received (1% Tween 80 in water), Standard group received Diclofenac-Na (25mg/Kg body weight i.p.), Group-I, Group-II and Group III were treated with 100, 200 and 400 mg/kg body weight of extract per oral. Values are mean ±SEM, (n = 5); ** $p < 0.001$, Dunnet test as compared to control

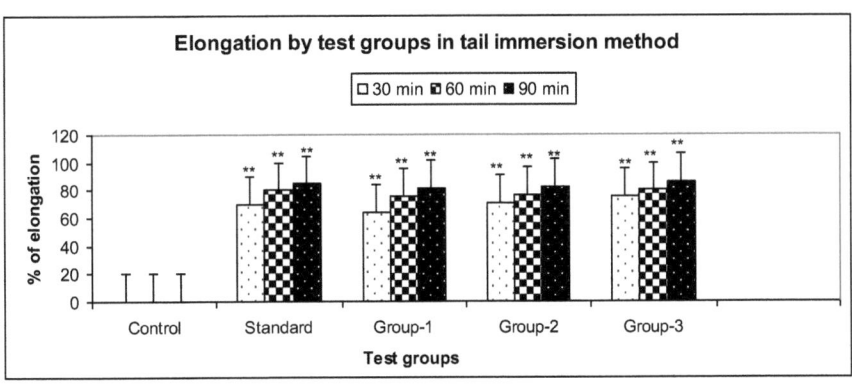

Control: animals received (1% Tween 80 in water), Standard group received Diclofenac-Na (25mg/Kg body weight i.p.), Group-I, Group-II and Group III were treated with 100, 200 and 400 mg/kg body weight of extract per oral. Values are mean ±SEM, (n = 5); ** $p < 0.001$, Dunnet test as compared to control

Figure 3.14: Percentage of elongation by *Hedychium coronarium in tail immersion method* (Dash et., 2011)

Discussion

The analgesic effect of the methanolic extract of the plant on acetic acid-induced writhing in mice was exhibited. The doses of the extract significantly ($p < 0.001$) inhibited writhing response induced by acetic acid in a dose dependent manner as compared to control. At 100 mg/kg body weight the extract showed 26.15% inhibition, at 200 mg/kg body weight the extract showed 47.94% inhibition and at 400 mg/kg body weight showed 73.12% inhibition of writhing compared to the standard drug Diclofenac- Na showed 75.78% inhibition of writhing at 50 mg/kg body weight dose. The tail withdrawal reflex time following administration of the extract of *Hedychium coronarium* was found to increase with increasing dose of the sample. In this test maximum effect was observed after 60 and 90 min of drug administration. The result was statistically significant ($p < 0.05$-0.001) and was comparable to the control. Acetic acid induced writhing test is suitable for detecting both central and peripheral analgesia, whereas tail flick tests are most sensitive to centrally acting analgesics. Intraperitoneal administration of acetic acid releases prostaglandins and sympathomimetic mediators like PGE_2 and $PGF_{2\alpha}$ and their levels increase in the peritoneal fluid of the acetic acid induced mice (Deraedt et al., 1980). The abdominal constrictions produced after administration of acetic acid is related to sensitization of nociceptive receptors to prostaglandins. It is therefore possible that the extract exerts its analgesic effect by inhibiting the synthesis or action of prostaglandins which may be due to phytochemicals present in the extract. Thermally induced nociception indicates narcotic involvement (Besra et al., 1996). The centrally acting analgesics generally elevate the pain threshold of mice towards heat. The extract significantly delayed the response time to thermal pain sensation in tail flick method indicating narcotic involvements. Moreover, since the extract inhibited both peripheral and central mechanisms of pain, it is possible that the extract acted on opioid receptor (Elisabetsky et al., 1995; Pal et al., 1999). Results of the present investigation suggest that the extract of *Hedychium coronarium* possesses strong analgesic activity and provide a scientific basis for the use of the plant in traditional system of medicine in the treatment of inflammatory disorders.

3.4.2 NEUROPHARMACOLOGICAl STUDY

CNS Depressant Activity Test of *Hedychium coronarium*

Drugs acting on the central nervous system (CNS) were first discovered by primitive humans and are still the most widely used group of pharmacologic agents CNS Action (Katzung, 1998). The effects of drugs on the central nervous system CNS with reference to the neurotransmitters for specific circuits, attenuation should be developed to general organizational principles of neurons. The view that synapses represent drug-modifiable control points within neuronal networks. It requires explicit delineation of the sites at which given neurotransmitters may operate and the degree of specificity with which such site that may be affected (Bloom, 1996).

Objectives

The purpose of this study was to examine neuropharmacological effect of methanolic extract of rhizomes of *Hedychium coronarium* on mice in a peripheral model of CNS depressant activity test.

Study Design

Experimental animals were randomly selected and divided into five groups denoted as group-I, group-II, group-III, group-IV, group-IV and group-V, consisting of 5 mice in each group. Each group received a particular treatment i.e. control, standard and the three doses of the extract. Each mouse was weighed properly and the doses of the test samples and control materials were adjusted accordingly.

Table 3.13: Experiment Profile to assess the effect of crude extract of *Hedychiumcoronarium* on CNS depressant activity test on Mice

Animal Group	Treatment	No. of Animals	Dose	Route of Administration
Control	Water (5ml/kg)	5	5 ml/kg	Oral
Standard	Diazepam	5	3 mg/kg	i.p
Group-I	Methanolic extract of *Hedychium coronarium* (100 mg/Kg dose)	5	100 mg/kg	Oral
Group-II	Methanolic extract of *Hedychium coronarium* (200 mg/Kg dose	5	200 mg/kg	Oral
Group-III	Methanolic extract of *Hedychium coronarium* (400 mg/Kg dose	5	400 mg/kg	Oral

Hole Cross Test

The most consistent behavioral change is a hyperemotional response to novel environmental stimuli. The aim of this study was to characterize the emotional behavior of mice using the hole-board test. The number of head-dips in the hole-board test in single-housed mice was significantly greater. Spontaneous movement of the animals through the hole from one chamber to the other was counted for 5 minuets in this test. The observations are made on 0, 30, 60, 90 and 120 minutes after orally administration of the test drugs.

Figure 3.15: Hole Cross Test

Preparation of the Test Materials

To prepare solution of the plant extract at a doses of 100mg/kg, 200mg/kg and 400mg/kg body weight, 180 mg extract was measured and added with it 9ml of distilled water and mixing with the help of vortex apparatus. From this solution 5ml/kg was taken for 100mg/kg, 200mg/kg and 400 mg/kg dose respectively.

Mechanism

The experiment was carried out as described by Takagi et al. (1971). Spontaneous movement of the animals through the hole from one chamber to the other was counted for 3 minuets in this test. The observations were made on 0, 30, 60, 90 and 120 minutes after orally administration of the rhizome extract of the *Hedychium coronarium*. There were no effects of the test animals at 0 min. After 30 min observed that the mice began to sleep and therefore very little movement was observed. Even after 120 min of administration of the extract they were still sleeping.

Results

Table 3.14: Effect of methanol extract of *Hedychium coronarium* on Hole cross test in mice. (Dash et al., 2011)

Group	Dose mg/kg	Number of movements % of Number of movements inhibition				
		0 min	30 min	60 min	90 min	120 min
Control		22.4±1.63	11.8±0.66	11.4±0.74	7.8±0.86	10.2±0.37
Standard	3	15.2±1.11	6.6±1.66* (44.07%)	4.0±1.09** (64.91%)	2.4±1.25* (69.23%)	1.6±0.87** (84.31%)
Group-I	100	14±2.74	9.4±1.63 (20.3%)	7.6±2.065 (33.33%)	4.6±1.53 (41.03)%	1.4±0.51** (86.24%)
Group-II	200	15±0.83	6.6±0.60* (44.07%)	5.2±0.20* (54.38%)	3.2±0.91* (58.97%)	2.0±0.54** (80.39%)
Group-III	400	14.2±4.3	7.2±2.22 (38.98%)	2.4±0.98** (78.95%)	1.4±0.74** (82.02%)	0.6±0.24** (94.12%)

Control: animals received (1% Tween 80 in water), Standard received Diazepam 3 mg/kg body weight i.p., Group-I, Group-II and Group III were treated with 100, 200 and 400 mg/kg body weight of the crude extract of *H. coronarium* per oral. Values are mean ±SEM, (n = 5); * $p < 0.05$, ** $p < 0.001$, Dunnet test as compared to control

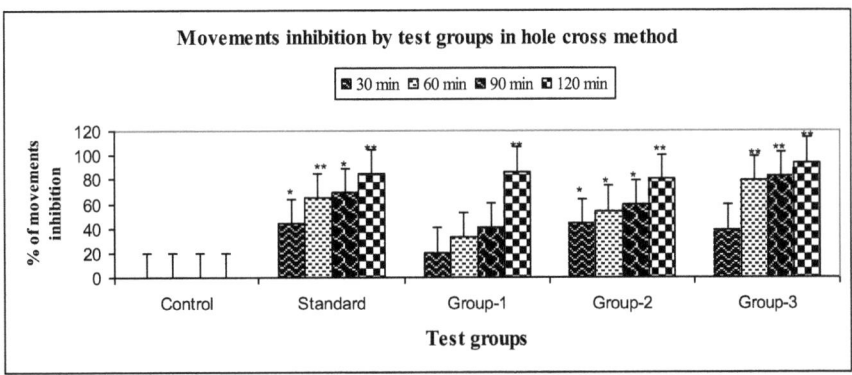

Control: animals received (1% Tween 80 in water), Standard received Diazepam 3 mg/kg body weight i.p., Group-I, Group-II and Group III were treated with 100, 200 and 400 mg/kg body weight of the crude extract of *H. coronarium* per oral. Values are mean ±SEM, (n = 5); * $p < 0.05$, ** $p < 0.001$, Dunnet test as compared to control.

Figure 3.16: Percentage of movements inhibitions by *Hedychium coronarium* in hole cross method (Dash et al., 2011)

Open Field Test

The Open Field Test (OFT) is clearly the most frequently used of all behavioural tests in pharmacology and neuroscience. Despite the simplicity of the apparatus, however, open field behaviour is complex. Consequently, it has been used to study a variety of behavioural traits, including general motor function, exploratory activity and anxiety-related behaviours. Open-field behavioral assays are commonly used to test both locomotor activity and emotionality in rodents Open Field Test for CNS Depressant Activity on Mice.

Preparation of the Test Materials

To prepare solution of the plant extract at a doses of 100mg/kg, 200mg/kg and 400mg/kg body weight, 180 mg extract was measured and added with it 9ml of distilled water and mixing with the help of vortex apparatus. From this solution 5ml/kg was taken for 100mg/kg, 200mg/kg and 400mg/kg dose respectively.

Mechanism

The method described by Gupta *et al.* (1971) was adopted for open field test. When the extract of the *Hedychium coronarium* were administered *at* the 0 min, there were no effects of the test animals. Within 60 min it was observed that the mice began to sleep and therefore little movement was observed. Even after 120 min of administration of the extract, they were still sleeping and there was no significant movement due to sleep.

Results

Table 3.15: Effects of methanol extract of *Hedychium coronarium* on open field test in mice. (Dash et al., 2011)

Group	Dose mg/kg	Number of movements (% of Number of movements inhibition)				
		0 min	30 min	60 min	90 min	120 min
Control		113 ±3.22	106.6±1.69 (-)	91.2 ±1.53 (-)	87.4 ±1.63 (-)	98±2.43 (-)
Standard	3	83.2±14.21	39.4±8.14** (63.03%)	32.6±6.22** (64.25%)	24.2±6.9** (72.31%)	11±3.115** (88.77%)
Group-I	100	52±6.26	30.2±3.12** (71.67%)	25.2±2.22** (72.37%)	22.6±2.27** (74.14%)	10±2.51** (89.79%)
Group-II	200	66.6±5.78	45.2±9.78** (57.60%)	27±7.98** (70.39%)	4.4±2.99** (94.96%)	2.2±1.02** (97.75%)
Group-III	400	51.4±14.46	30.6±3.4** (71.29%)	19.2±0.735** (78.94%)	10.2±1.67** (88.32%)	3.4±0.745** (96.53%)

Control: animals received (1% Tween 80 in water), Standard received Diazepam 3 mg/kg body weight, Group-I, Group-II and Group III were treated with 100, 200 and 400 mg/kg body weight of the crude extract of *H. coronarium*. Values are mean ±SEM, (n = 5); ** $p < 0.001$, Dunnet test as compared to control

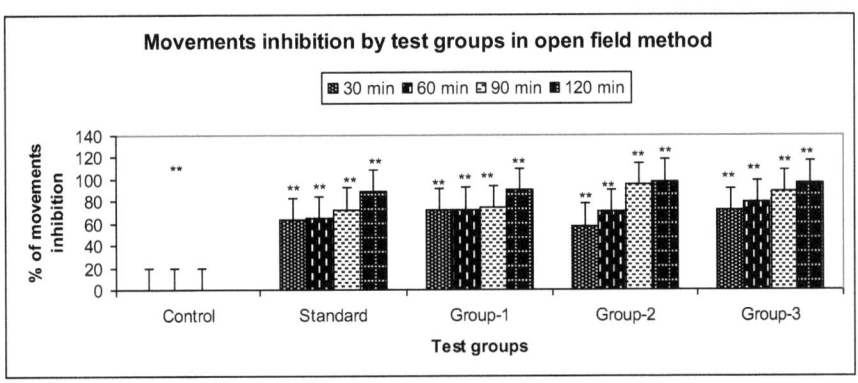

Control: animals received (1% Tween 80 in water), Standard received Diazepam 3 mg/kg body weight, Group-I, Group-II and Group III were treated with 100, 200 and 400 mg/kg body weight of the crude extract of *H. coronarium.* Values are mean ±SEM, (n = 5); ** $p < 0.001$, Dunnet test as compared to control

Figure 3.17: Percentage of movements inhibitions by *Hedychium coronarium* in open field method (Dash et al., 2011)

Discussion

The extract at doses level of 100mg/kg, 200mg/kg and 400mg/kg body weight showed significant ($p < 0.001$) decrease of movement from its initial value during the period of hole cross experiment as compared to control. The maximum decrease in movement was observed at 90 and 120 min after drug administration. In the open field test at dose level of 100mg/kg, 200mg/kg and 400mg/kg body weight the number of squares traveled by the mice was suppressed significantly from its initial score by both doses of the extract which is comparable to the standard drug Diazepam. The maximum suppression was exhibited at 90 and 120 min after drug administration. While evaluating neuropharmacological activities of *Hedychium coronarium,* it was found that the plant extract possesses central nervous system depressant activity as indicated by decreased exploratory behaviour in mice. The extract also displayed a marked sedative effect as indicated by the reduction in gross behaviour and potentiation of thiopental sodium induced sleeping time. It is generally accepted that the sedative effect of drugs can be evaluated by measurement of spontaneous motor activity and barbiturate induced sleeping time in laboratory animal model (Lu, 1998). This result confirms those of Fujimori (1995) who proposed that the enhancement of barbiturate induced hypnosis is a good index of CNS depressant activity.

The most important step in evaluating drug action on CNS is to observe its effect on locomotors activity of the animal. The activity is a measure of the level of excitability of the CNS (Mansur, *et al.,* 1980) and this decrease may be closely related to sedation resulting from depression of the central nervous system. The extract significantly decreased the locomotor activity as shown by the results of the hole cross and open field tests. The locomotor activity lowering effect was evident at the 3rd observation (60 min) and continued up to 5th observation period (120 min). Open field test showed that the depressing action of the extracts was evident from the second observation period in the test animals at the dose of 100mg/kg, 200 mg/kg and 400 mg/kg body weight. Maximum depressant effect was observed from 2nd (30 min) to 5th (120 min) observation period at the dose of 100mg/kg, 200 mg/kg and 400 mg/kg. The results were also dose dependent. Results of the present investigation suggest that the extract of *Hedychium coronarium* possesses strong CNS depressant activity and provide a scientific basis for the use of the plant in traditional system of medicine in the treatment of inflammatory disorders

3.4.3 CYTOTOXICITY ACTIVITY TEST

Brine Shrimp Lethality Bioassay

Bioactive compounds are almost always toxic in high doses. Pharmacology is simply toxicology at a lower dose and toxicology is simply pharmacology at a higher dose. Brine Shrimp Lethality Bioassay is a bench top bioassay method for evaluating anticancer, antimicrobial and pharmacological activities of natural products. It is a recent development in the field of bioassay for bioactive compounds. By this method natural product extracts, fractions as well as pure compounds can be tested for their bioactivity. Here in vivo lethality, a simple zoologic organism is used as convenient monitor for screening and in the discovery of new bioactive natural products. The bioassay is indicative of cytotoxicity and a wide range of pharmacological activity of the compounds.(Mc Laughlin,1990; Persoone 1980; Meyer et al.,1982). Here the simple zoologic organism is brine shrimp nauplii. The eggs of brine shrimp, *Artemia salina* (leach) are readily available at low cost and remain viable for year in the dry state. Upon being placed in seawater, the egg hatch within 48 hours to provide a large number of larvae (nauplii) for experimental uses.

Principle

Brine Shrimp lethality bioassay is a rapid and comprehensive bioassay for the bioactive compounds of natural and synthetic origin. By this method, natural product extracts, fractions as well as the pure compounds can be tested for their bioactivity. The method utilizes in vivo lethality in a simple zoological organism (Brine nauplii) as a convenient monitor for screening and fractional in the discovery of new bioactive natural products. Brine toxicity is closely correlated with 9KB (human nasopharyngeal carcinoma) cytotoxicity (p=0.036 and kappa = 0.56). ED_{50} values for cytotoxities are generally about one-tenth the LC_{50} values found in the Brine Shrimp test. Thus, it is possible to detect and then monitor the fractional of cytotoxicity, as well as 3PS (P_{388}) (In vivo murine leukaemia) active extracts using the Brine lethality bioassay. The Brine Shrimp assay has advantages of being rapid (24 hours), inexpensive, and simple (e.g., no aseptic technique are required). It easily utilizes a large number of organisms for statistical validation and requires no special equipment and a relatively small amount of sample (2-20 mg or less). Furthermore, it does not require animal serum as in needed for cytotoxicities.

Materials

- *Artemia salina* Leach. (Brine eggs), Sea salt (Nacl)
- Dimethyl sulfoxide (DMSO)
- Small tank with perforated dividing dam to hatch the Shrimp
- Lamp to attract Shrimps
- Pipettes (5, 25ml) and Micropipette (100-1000µl)
- Glass vials, Magnifying glass

Preparation of simulated seawater

38g of sea salt (pure NaCl) was weighed by rough balance and dissolved in one liter of distilled water in a small tank and then filtered off to get clear solution. This simulated seawater was for hatching of brine shrimp.

Hatching of Brine Shrimps

Artemia saline leach (Brine Shrimp eggs) collected from pet shops was used as the test organism. Simulated seawater was taken in a small tank and Shrimp eggs were added to the one side of the perforated divided tank and then this side was covered. The tank was kept under constant aeration for 48 hrs to hatch the shrimp and to be matured as nauplii (larvae).The hatched Shrimps were attracted to the lamp through the perforated dam and with the help of a pasture pipette. 10 living Shrimps were added to each of the test tubes containing 5 ml of Brine solution.

Figure 3.18 : Hatching of Brine Shrimps (Dash et al., 2013)

Preparation of Test Solutions

5 mg of each of the extracts was measured and dissolved in DMSO. Finally the concentration was adjusted to 320μg/ml that served as the mother solution. A series of solutions of lower concentrations were prepared by serial dilution with DMSO. From each of these test solutions 30μl were added to pre-marked glass vials/test tubes containing 5ml of seawater and 10 shrimp nauplii. So, the final concentration of samples in the Vials/test tubes were 320μg/ml, 160μg/ml, 80μg/ml, 40μg/ml, 20μg/ml, 10μg/ml, 5μg/ml, 2.5μg/ml, and 1.25μg/ml.

Preparation of Positive control

Vincristine sulphate served as the positive control. 0.2mg of Vincristine sulphate was dissolved in DMSO to get an initial concentration of 20μg/ml from which serial dilutions were made using DMSO to get 10μg/ml, 5 μg/ml, 2.5 μg/ml, 1.25 μg/ml, 0.625 μg/ml, 0.3125 μg/ml, 0.15625 μg/ml, 0.078125 μg/ml, 0.0390 μg/ml.The control groups containing 10 living Brine Shrimp nauplii in 5ml simulated seawater received the positive control solutions.

Preparation of Negative control

As for negative control, 30μl of DMSO was added to each of the pre-marked test tubes containing 5ml of simulated seawater and 10 shrimp nauplii. The test was considered invalid if the negative control showed a rapid mortality rate and therefore conducted again. The test tubes (containing nauplii) were then maintained at room temperature for 24 hrs under the light for observing the survival rate.

Counting of Nauplii and Analysis of Data

After 24 hours, the test tubes were inspected using a magnifying glass and the number of survivors was counted. The percent (%) mortality was calculated for each dilution. The concentration-mortality data were analyzed by using probit analysis and linear regression. The effectiveness or the concentration-mortality relationship of plant product is usually expressed as a median lethal concentration (LC_{50}) value. This represents the concentration of the chemical that produces death in half of the test subjects after a certain exposure period.

Results and Discussion

Following the procedure of Meyer (Meyer et al., 1982), the methanolic extract showed lethality indicating the biological activity of the compound present in the extract. Test sample showed different mortality rate at different concentrations. The mortality rate of brine shrimp was found to be increased in concentration of the sample and plot of percent mortality versus Log concentration on the graph paper produced an approximate linear correlation between them. From the graph (Figure) the concentration at which 50% mortality (LC_{50}) of brine shrimp nauplii occurred were obtained by extrapolation. The values were found to be 0.39µg/ml for crude extract. The same samples were tested for Brine Shrimp Lethality Bioassay in four times and the average data is given below:

Table 3.16: Effect of *Hedychium coronarium* on brine shrimp lethality test in *Artemia salina*.(Dash et al., 2015)

conc. (µg/ml)	Log conc.	No.of nauplii taken	Trial-1	Trial-2	Trial-3	Trial-4	Average no. of nauplii dead	% mortality	Std.Conc (µg/ml)	Log conc.	Nauplii taken	Nauplii dead	% mortality
1.25	0.090	10	5	3	4	4	4	40	0.156	-0.806	10	3	30
2.5	0.39	10	6	5	5	4	5	50	0.312	-0.505	10	4	40
5	0.69	10	6	6	6	5	5.75	57.5	0.625	-0.204	10	5	50
10	1	10	7	7	6	5	6.25	62.5	1.25	0.096	10	6	60
20	1.30	10	7	7	6	6	6.5	65	2.5	0.397	10	8	80
40	1.60	10	8	7	7	6	7	70	5	0.698	10	9	90
80	1.90	10	8	7	7	6	7	70	10	1	10	10	100
160	2.20	10	8	8	7	7	7.5	75	20	1.310	10	10	100
320	2.50	10	9	9	7	7	8	80	40	1.602	10	10	100

Mortality (%) = Number of dead brine shrimps × 100/ Total number of brine shrimps.

Table 3.17: Result of *Hedychium coronarium* against on *Artemia salina*. (Dash et al., 2015)

Sample	LC$_{50}$ (µg/ml)	Regression equation	R^2
Vincristine Sulphate	0.52	y = 32.61x+59.22	0.942
Methanolic extract	0.39	y = 14.81x+44.065	0.944

The lethal concentration LC$_{50}$ of the test samples after 24 hr. was obtained by a plot of percentage of the shrimps killed against the logarithm of the sample concentration (toxicant concentration) and the best fit line was obtained from the curve data by means of regression analysis.

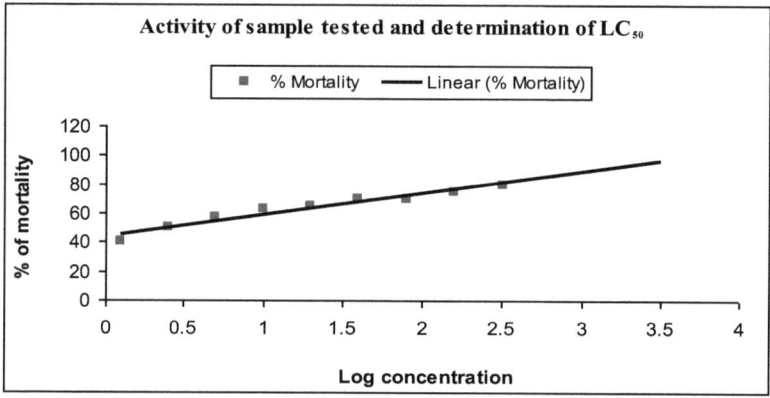

Figure 3.19: Effect of methanolic extract of *Hedychium coronarium* on Brine shrimp nauplii. (Dash et al., 2015)

The degree of lethality to *Hedychium coronarium* was observed with exposure to different dose different dose levels of the test samples. The degree of lethality was directly proportional to the concentration of the extract ranging from significantly with the lowest concentration (1.25µg/ml) to highly significant with the highest concentration (320µg/ml).Maximum mortalities took place from the concentration 320µg/ml, whereas least mortalities were at 1.25 µg/ml concentration. In other words, mortality increased gradually with the increase in concentration of the test samples.

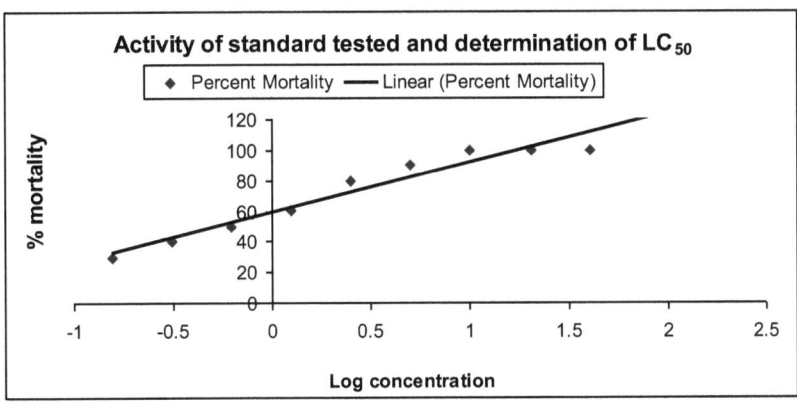

Figure 3.20: Effect of Vincristine Sulphate *on* Brine shrimp nauplii (Dash et al., 2015)

Comparison with positive control Vincristine and negative control signifies that cytotoxicity exhibited by the crude extract is promising and further bioactivity-guided investigation can be done to find out the active potent antitumor and pesticidal compounds. In the results of bioassay, test sample were found to show significant activity against brine shrimp nauplii. Since the chemical constituents present in a plant are directly responsible for its therapeutic and other pharmacological properties, the constituents of the plant which were reported and /or dected during the phytochemical investigations should have some direct relationship with local medicinal uses. Thus it was expected that the selected plant analyzed under this research work should have some therapeutic effects. The results of phytochemical test and surveyed literature revealed that the leaves and stems contain a number of chemical substances of which some might be toxic. Thus it was assumed that the plant extracts, analyzed under this work, might have the expected effects and/or other therapeutic effects. (Ghani, 1998). Alkaloids of the plant might be responsible for cytotoxic effect. When these constituents are present in a plant, probably the plant show bioactivity. Although after 24 hours the samples showed significant lethality, it was not accepted for the safest and more accurate result. Because after 24 hours some nauplii may die normally as their life span is 24 to 48 hours, though it was not seen in the table of lethality that after 24 hours too, in the control and in the 0µg/ml concentration vials, at least more than 50% of nauplii were alive. So it was proved that the control sample i.e. the only seawater and zero concentration sample i.e. the mixture of DMSO and sea water solution were not toxic. The toxicity that found in other vials of different concentrations was only for the toxic property of sample i.e. is the extract of the plant. So it was proved that the sample under test had effective toxic property. Thus it was assumed that the sample might have good cytotoxic property. Many of the cytotoxic drugs used in cancer chemotherapy have been observed to produce immunosuppression, leading to investigation of the usefulness of these compounds immunosuppressive agents in that they prevent the clonal expansion of both B and T lymphocytes (Hardman *et al.,* 1996). As methanol extract possess cytotoxic property it may be used as immunosuppressive agents in the prevention of transplant rejection and the treatment of autoimmune disorders. However, further researches are necessary particularly with its purified fraction

CHAPTER 4: CONCLUDING REMARKS

Conclusion

Plants are most valuable sources of biologically active products, which could provide main chemical constituents, which are of prime importance in the fight against diseases, pain, infections and ultimately death. The plant *Hedychium coronarium* has many pharmacological properties, which are directly related to the chemical constituents of the plant. Thus, it was expected that the selected plant analyzed under this research work should have some direct therapeutic effects. In light of the results of the present study, it can be concluded that the methanolic rhizome extract possesses remarkable analgesic, CNS depressant and cytotoxic potential. The findings of the investigation also provide further support to and reinforce the traditional use of the plant in different medical disorders. Positive result in cytotoxic activity tests led us to the inference that the plant extract may contain bioactive compounds which may aid ongoing anticancer drug discovery from floristic resources. The results of phytochemical screening showed that the methanolic rhizome extract of *Hedychium coronarium* contained carbohydrates, flavonoids, saponins, steroids and alkaloids. These compounds are responsible for the bioactivities but it is quit difficult to ascribe the observed activities to any specific group of compounds. Hence, further studies are suggested to be undertaken to pinpoint the compounds found in the methanolic rhizome extract of *Hedychium coronarium* and to better understand the mechanism of such actions scientifically.

CHAPTER 5: REFERENCES

Adams RP (2001). Identification of essential oil components by gas chromatography/mass spectroscopy, Allured Publishing, Carol Stream. p. 456.

Akiyama K, Kikuzaki H, Aoki T, Okuda A, Lajis NH, Nakatani N (2006). *J. Nat. Prod*, 69: 1637—1640

Aziz MA, Habib MR, Karim MR (2009). Antibacterial and cytotoxic activity of *Hedychium coronarium* J. Koeing: *Research Journal of Agriculture and Biological Sciences*, 5 (6):969-972

Ampofo O (1979). In African Medicinal Plants, (ed. Sofowara), University of life Negeria.

Bhandary MJ, Chandrashekar KR, Kaveriappa KM (1995). Medical ethnobotany of the Siddis of Uttara Kannada district, Karnataka, India. *J. Ethnopharmacol*, 47: 149-158

Bingel, A.S. and Farnsworth, N.R., (1960), *Journal of American Medical Society*, 5: 25-26.

Barel S, Segal R, Yashphe J (1991). The antimicrobial activity of the essential oil from Achillea fragrantissima. *J Ethnopharmacol*, 33: 187–191.

Beena Joy, Ranjan A, Abraham E (2007). Antimicrobial activity and chemical composition of essential oil from *Hedychium coronarium*: *Phytother. Res*, 21: 439-443.

Bamgbose SO, Noamesi BK (1981). Studies on cryptolepine. II: Inhibition of carrageenan- induced oedema by cryptolepine. *Planta Med*, 41: 392 396.

Beena J, Akhila R, Emilia A (2007). Antimicrobial activity and chemical composition of essential oil from *Hedychium coronarium*. *Phytother. Res*, 21: 439-443.

Bisht GS, Awasthi AK, Dhole TN, (2006). Antimicrobial activity of *Hedychium spicatum*. *Phytother. Res*, 21: 439-443.

Bunzel M, Ralph J, Funk C, Steinhart H (2005).*Tetrahedron Lett*, 46, 5845—5850.

Bonjar SGH, (2004). Evaluation of antibacterial properties of Iranian medicinal-plants against *Micrococcus luteus*, *Serratia marcescens*, *Klebsiella pneumoniae* and *Bordetella bronchoseptica*. *Asian J. Plant Sci*, 3: 82-86.

Cardador-Martinez A, Loarca-Pina G, Oomal BD (2002). Antioxidant activity in common beans (*Phaseolus vulgaris* L.). *J. Agric. Food Chem*, 50: 6975-6980.

Crunkhon P, Meacock SER (1971). Mediators of the inflammation induced in the rat paw by carrageenan. *Br. J. Pharmacol*, 42: 392-402.

Chandoke, N. and Rayghatak, B. J.(1969). Studies on *Tagetes minuta: Some* Pharmacological actions of essential oil.*Ind. Jour. Med. Res.*, 57, 864-876.

Cheong H, Choi E J, Yoo GS, Kim KM, Ryu SY(1998). *Planta Med*, 64:577-578.

Chopra RN, Nayar SI, Chopra LE (1956).Glossary of Indian Medicinal Plants. Council of Scientific and Industrial Research, p57.

Chopra RN, Chopra IC, Verma BS (1969). *Supplement to Glossary of Indian Medicinal Plants*. New Delhi: Publication and Information Directorate. p. 35.

Chimnoi NS, Pisutiaroenpong L, Ngiwsara D, Dechtrirut D, Chokchaichamnanki TN, Khunnawutmanotham C, Mahidol, Techasakul S, (2008). Labdane diterpenes from the rhizomes of *Hedychium coronarium*. Nat. Prod. Res, 22(14): 1255-1262.

Crunkhorn P, Meacock SC (1971). Mediators of the inflammation induced in the rat paw by carrageenin. *Br. J. Pharmacol*, 42: 392-402.

Caccioni DRL, Guizzardi M, Biondi DM, Renda A Ruberto G. 1998. Relationship between volatile components of citrus fruit essential oils and antimicrobial action on Penicillium digitatum and Penicillium italicum. Int J Food Microbiol 43: 73–79.

Carson CF, Riley TV (1995). Antimicrobial activity of the major components of the essential oil of Melaleuca alternifolia. *J Appl Bacteriol,* 74: 264–269.

Chen JJ, Ting CW, Wu YC, Hwang TL, Cheng TL, Cheng MJ, Sung PJ, Wanf TC, Chen JF (2013). New labdane –type Diterpenoids and anti-inflammatory constituents from Hedychium coronarium. *Int. J. Mol. Sci*, 14: 13063-13077.

Cosentino S, Tuberoso CIG, Pisano B et al (1999). In vitro antimicrobial activity and chemical composition of Sardinian Thymus essential oils. *Lett Appl Microbiol* 29: 130–135.

D'Amour FE, Smith DL (1941). A method for determining loss of pain sensation. *J. Pharmacol. Exp. Ther*, 72: 74-79.

Di Rosa M and Willoughby DA (1971). Screens for anti-inflammatory drugs. *J. Pharm. Pharmacol.*, 23: 297-298.

Daferera DJ, Ziogas BN, Polissiou MG (2000). GC-MS analysis of essential oils of some Greek aromatic plants and their fungi toxicity on Penicillium digitatum. *J Agric Food Chem* 48: 2576–2581.

Davies NW (1990). Gas chromatographic retention indices of monoterpenes and sesquiterpenes on methylsilicone and Carbowax 20M phases. J. Chromatogr. 503: 1-24.

Dixit VK, Varma KC, (1979). Effects of essential oils of rhizomes of *Hedychium coronarium* and *Hedychium spicatum* on Central Nervous System: *Ind. J. Pharmal*, 2(2): 147-149.

Dash PR, Nasrin M and Saha MR (2011). Evaluation of Analgesic and Neuropharmacological activities of methanolic rhizome extract of *Hedychium coronarium*. *International Journal of Pharmaceutical Sciences and Research*, **2**(4): 979-984.

Dash PR and Sheikh Z (2015). Preliminary studies on phytochemicals and cytotoxic activity of methanolic rhizome extract of *Hedychium coronarium*. *Journal of Pharmacognosy and Phytochemistry,* 4(1): 136-139.

Dash PR, Raihan SZ and Ali MS (2013). Ethnopharmacological investigation of the spice Kaempferia galangal. Lambert Academic Publishing, 1st edition, German

Dandiya PC, Cullumbine H (1959). Studies on *Acorus calmus:* Pharmacological actions of essential oil. *J. Pharmac. exp. Ther,* 125. 353-359.

Dandiya PC, and Menon MK (1963). Central Nervous System Depress sants III: Influence of some tranquillizing agents on morphine analgesia. *Arch. int. Pharmacodyn,* 141: 223-232

Finney DJ, (1971). Probit Analysis. Cambridge University Press. London, pp: 333.

Gupta SS (1994). Prospects and perspectives of natural plant products in medicine. *Indian J Pharmacol,* 26: 1–12.

Gopanraj G, Dan M, Shiburaj S, Sethuraman MG, George V (2005).Chemical composition and antibacterial activity of the rhizome oil of *Hedychium larsenii*. *Acta Pharm*. 55: 315-320.

Ghani A, (1998). Medicinal Plants of Bangladesh with Chemical Constituents and Uses. 2nd edition. Asiatic Society of Bangladesh, Dhaka, pp.4-19.

Ghani A, (2005). Text Book of Pharmacognosy, 2nd Edition, pp. 197-205.

Ghani A, (2005). Practical Phytochemistry, pp. 149-152 Parash Publishers.

Garden Plants. *Hedychium coronarium* Koenig. Civil Service Development Institute, Taipei, Taiwan.

Godofredo U and Stuart Jr (2016). Philippine Medicinal Plants. Kamia.

Harborne JB, (1984). Phytochemical methods, Champmann and Hall Ltd. London, pp: 134.

Hedichium coronarium, lirio-do-brejo.

Itokawa H, Morita H, Katou I, Takeya K, Cavalheiro AJ, Oliveira RCB, Ishige M, Motidome M (1988a). Cytotoxic diterpenes from the rhizomes of *Hedychium coronarium*. *Planta Med*, 55: 311-315.

Itokawa H, Morita H, Takeya K, Motidome M (1988b). Diterpenes from rhizomes of *Hedychium coronarium*. *Chem. Pharm. Bull*, 36: 2682-2684.

Itokawa H, Morita H, Katou I, Takeya K, Cavalheiro AJ, De Oliveira RCB, Ishige M, Motidome M(1988). *Planta Med*, 54: 311-315.

Joy B, Rajan A, Abraham E (2007). Antimicrobial activity and chemical composition of essential oil from *Hedychium coronarium*. *Phutother. Res*, 21: 439-443.

Jain SK, Fernandes VF, Lata S, Ayub A (1995). Indo-Amazonian ethnobotanic connections Similar uses of some common plants.Ethnobotany, 7: 29-37.

Jadhav V, Kore A, Kadam VJ (2007). In-vitro pediculicidal activity of *Hedychium spicatum* essential oil. *Fitoterapia*, 78: 470-472.

Jansen AM, Scheffer JJ, Svemdsen B (1987). Antimicrobial activity of essential oils: A 1976–1986 literature review: aspects of test methods. *Planta Med,* 53: 395–398.

Jirovetz L, Buchbauer G, Denkova Z (2005). Antimicrobial testings and gas chromatographic analysis of pure oxygenated monoterpenes 1,8-cineole, α-terpineol, terpinen-4-ol and camphor as well as target compounds in essential oils of pine (Pinus pinaster), rosemary (Rosmarinus officinalis), tea tree (Melaleuca alternifolia). *Sci Pharm* 73: 27–39.

Kiritikar KR, Basu BD, (1933). Indian Medicinal Plants. Lalit Mohan Basu, Allahabad, pp: 1478-1480.

Kandil O, Radwan NM, Hassan AB, Amer AMM, El-Banna HA, Amer WMM (1994). Extracts and fractions of Thymus capitatus exhibit antimicrobial activities. *J Ethnopharmacol* 44: 19–24.

Kiritikar KR, Basu BD, (1933). In Indian medicinal plants, Lalit Mohan Basu, Allahabad, vol II, 2nd edn, pp.1478–1480.

Li N, Morikawa T, Ninomiya K, Li X., Yoshikawa M, (2007). *Heterocycles*, 71: 1193-1201

Lu Y, Zhong CX, Lu C, Li XL, Wang PJ (2009). Anti-inflammation activity and chemical composition of flower essential oil from *Hedychium coronarium*. *African Journal of Biotechnology*. 8 (20): 5373-5377.

Morikawa T, Matsuda H., Sakamoto Y, Ueda K, Yoshikawa M (2002). *Chem. Pharm. Bull*, 50: 1045-1049.

Matsuda H, Morikawa T, Sakamoto Y, Toguchida I., Yoshikawa M (2002). *Bioorg. Med. Chem*, 10: 2527-2534.

Morikawa T, Li X, Nishida E, Ito Y, Matsuda H, Nakamura S, Muraoka O, Yoshikawa M (2008). *J. Nat. Prod,* 71: 828-835.

Matsuda H, Ninomiya K, Morikawa T, Yasuda D, Yamaguchi I., Yoshikawa M (2008). *Bioorg. Med. Chem. Lett.*, 18: 2038-2042.

Matsuda H, Morikawa T, Sakamoto Y, Toguchida I., Yoshikawa M (2002). *Heterocycles*, 56: 45-50

Mishra M (2013). Current status of endangered medicinal plant *Hedychium coronarium* and causes of population decline in the natural forests of Anuppur and Dindori districts of Madhya Pradesh, India. *Int. Res. J. of Bio. Sci*, 2(3): 1-6.

Matsuda H, Morikawa T, Ninomiya K, Yoshikawa M (2001). *Bioorg Med. Chem*, 9:909-916.

Matsuda H., Morikawa T, Toguchida I, Ninomiya K., Yoshikawa M (2001). *Heterocycles*, 55: 841-846.

Matsuda H., Morikawa T, Ninomiya K, Yoshikawa M (2001).*Tetrahedron*, 57:8443—8453.

Muraoka O, Fujimoto M, Tanabe G, Kubo M, Minematsu T, Matsuda H, Morikawa T, Toguchida I, Yoshikawa M (2001). *Bioorg. Med. Chem. Lett*, 11: 2217-2220.

Matsuda H, Morikawa T, Toguchida I, Ninomiya K, Yoshikawa M, (2001). *Chem. Pharm. Bull.*, 49:1558—1566.

Matsuda H, Morikawa M, Tao J, Ueda K, Yoshikawa M (2002). *Chem. Pharm. Bull.*, 50: 208-215.

Morikawa T, Matsuda H, Ninomiya K, Yoshikawa M, (2002). *Biol. Pharm. Bull.*, 25; 627-631.

Morikawa T, Matsuda H, Ninomiya K, Yoshikawa M (2002a). Medicinal foodstuffs. XXIX. Potent protective effects of sesquiterpenes and curcumin from Zedoariae Rhizoma on liver injury induced by Dgalactosamine/ lipopolysaccharide or tumor necrosis factor-alpha. *Biol. Pharm. Bull*. 25: 627-631.

Morikawa T, Matsuda H, Sakamoto Y, Ueda K, Yoshikawa M (2002b). New farnesane-type sesquiterpenes, hedychiols A and B 8, 9- diacetate, and inhibitors of degranulation in RBL-2H3 cells from the rhizome of *Hedychium coronarium*. *Chem. Pharm. Bull*. 50: 1045- 1049.

Muraoka O, Fujimoto M, Tanabe G, Kubo M, Minematsu T, Matsuda H, Morikawa T, Toguchida I, Yoshikawa M (2001). Absolute stereostructures of novel norcadinane- and trinoreudesmane-type sesquiterpenes with nitric oxide production inhibitory activity from *Alpinia oxyphylla*. *Bioorg. Med. Chem. Lett.* 11: 2217-2220.

Nakatani N, Kikuzaki H, Yamaji H, Yoshio K, Kitora C, Okada K, Padolina WG (1994). Labdane diterpenes from rhizomes of *Hedychium coronarium*. *Phytochemistry*, 37: 1383-1388.

NIH (1985). Guide for the use of laboratory animals. DHHS, PHS, NIH Publication No. 85-23.

Nakatsu T, Lupo AT, Chinn JW, Kang RKL (2000). Biological activity of essential oils and their constituents. Stud. Nat. Prod. Chem. **21** (Bioactive Natural Products (Part B)), 571– 631, Elsevier.

Nakatani N, Kikuzaki H, Yamaji H, Yoshio K, Kitora C, Okada K and Padolina WG (1994). Labdane diterpenes from rhizomes of *Hedychium coronarium*. *Phytochemistry*, 37: 1383-1388.

Nakamura S, Li X, Matsuda H, Ninomiya K, Morikawa T, Yamaguti K., Yoshikawa M (2007). *Chem. Pharm. Bull.*, 55: 1505—1511.

Nakamura S, Sugimoto S, Matsuda H, Yoshikawa M (2007). *Chem. Pharm. Bull*, 55:1342-1348.

Nakamura S, Okazaki Y, Nonomiya K, Morikawa T, Matsuda H, Yoshikawa (2008).*Chem Pharm. Bull,* 56(12):1704-1709.

Ninomiya K, Morikawa T, Zhang Y, Nakamura S, Matsuda H, Muraoka O, Yoshikawa M (2007). , *Chem. Pharm. Bul*, 55: 1185—1191.

Oka M, Maeda S, Koga N, Kato K, Saito T(1992). *Biosci. Biotech. Biochem*, 56: 1472-1473.

Ohtani I., Kusumi T., Kashman Y., Kakisawa H., *J. Am. Chem. Soc.*, **113**, 4092—4096 (1991).

Oloke JK, Kolawole BO, Erhun WO (1988). The antibacterial and antifungal activities of certain components of Aframomum melegueta. *Fitoterapia* 59: 384–388.

Porter NG, Wilkins AL (1999). Chemical, physical and antimicrobial properties of essential oils of Leptospermum scoparium and Kunzea ericoides. *Phytochemistry* 50: 407–415.

Xie Y., Morikawa T, Ninomiya K, Imura K, Muraoka O, Yuan D, Yoshikawa M (2008). *Chem. Pharm. Bull*, 56: 1628—1631.

Rois JJ, Reico MC, Villar A (1988). Antimicrobial Screening of natural products. *J. Enthopharmocol*, 23: 127-149.

Reiner R., (1982). Detection of antibiotic activity. In Antibiotic and Introduction, Roche Scientific Services, Switzerland, pp: 21-27.

Ribeiro RA, Fiuza de Melo MM, De Barros F, Gomes C, Trolin G, (1986). Acute antihypertensive effect in conscious rats produced by some medicinal plants used in the sate of São Paulo. *J. Ethnopharmacol*, 15(3): 261-269.

Ramarao AV, Gurjar MK (1990). Drugs from plant resources: an overview. Pharma Times, 22(5): 19-21.

Ribeiro RA, De Barros F, Fiuza de Melo MM, Muniz C, Chieia S, Wanderley MG, Gomes C, Trolin G, (1988). Acute diuretics effects in conscious rats produced by some medicinal plants used in the sate of São Paulo. *J. Ethnopharmacol*, 24(1): 19-29.

Soares DJ, Barreto RW (2008). Fungal pathogens of the invasive riparian weed *Hedychium coronarium* from Brazil and their potential for biological control. *Fungal Diversity* 28: 85-96.

Sabual and Baby(2007). Chemical composition and antimicrobial activities of the essential oils from the rhizomes of four Hedychium sp. from South India. *Journal of Essential oil Research*.

Sing DL, Sing LR, Devi PG, Devi NR, Sing LS, Bag GC (2013): Comparative study of phytochemical constituents and total phenolic conent in the extract of three different species of genus Hedychium. *Int. Journal of Pharma Tech. Research*, 5(2): 601-606.

Schwartz LB, Lewis RA, Seldin D, Austen KF (1981). *J. Immunol*, 126:1290-1294.

Shrotriya S, Ali MS, Saha A, Bachar SC, Islam MS, (2007). Anti-inflammatory and analgesic effects of *Hedychium coronarium* Koen. *Pak. J. Pharm. Sci*, 20(1): 47-51.

Shrotriya S, Ali MS, Saha A, Bachar SC, Islam MS, (2007).. Anti-inflammatory and analgesic effects of *Hedychium coronarium* Koen. *Pak. J. Pharm. Sci*, 20(1): 42-47.

Sofowara A (1982). Medicinal Plants and Traditional Medicine in Africa, John Wiley and Sons Ltd. New York, USA, pp.54.

Sosa S, Balicet MJ, Arvigo R, Esposito RG, Pizza C and Altinier GA (2002). Screening of the topical anti-inflammatory activity of some Central American Plants. *J. Ethanopharmacol*, 8: 211-215.

Smid EJ, Gorris LGM (1999). Natural antimicrobials for food preservation. In Handbook of Food Preservation, Rahman S (ed.). Marcel Dekker: New York, 285–308.

Sushil J, Chandan SC, Garima A, Om P, Anil KP, Chandra SM (2008). Terpenoid compositions, and antioxidant and antimicrobial properties of the rhizome essential oils of different *Hedychium* Species. *Chem. Biodivers*, 5: 299-309.

Seikou N, Yoshie O, Kiyofumi N, Toshio M, Hisashi M, Masayuki Y (2008). Medicinal Flowers. XXIV. 1) Chemical structures and hepatoprotective effects of constituents from flowers of *Hedychium coronarium*. *Chem. Pharm. Bull*, 56: 1704-1709.

Thanh BV, Dai DN, Thang TD, Binh NQ, Anh LDN, Ogunwande IA (2014). Composition of essential oils of four Hedychium species from Vietnam. *Chemistry Central Journal*, 8:54.

Vaidya AB, Antarkar VDS. (1994). New drugs from medicinal plants: opportunities and approaches. *J Assoc Physicians India* 42: 221–228.

Verma M, Bansal YK (2012). Induction of somatic embroyogenesis in endangered butterfly ginger *Hedychium coronarium* J. Koenig, *Indian Journal of Exp. Biology*, 50: 904-909.

Verma M, Bansal YK (2013). Effect of additives on plant regeneration in Hedychium coronarium J. Koenig an endangered aromatic and medicinal herb. *Int. J. Pharm. Sci. Rev. Res*, 23(1):105-110.

Viljoen A, van Vuuren S, Ernst E et al. (2003). Osmitopsis asteriscoides (Asteraceae) – the antimicrobial activity and essential oil composition of a Cape-Dutch remedy. J Ethnopharmacol, 88: 137–143.

Vinegar R, Schreiber W, Hugo R (1969). Biphasic development of carrageenin oedema in rats. *J. Pharmacol. Exp. Ther,* 166: 96-103.

Vogel HG, Vogel WH, (1997). Pharmacological Assays. *In:* Drug Discovery and Evaluation: Springer Verlag, Germany, pp.368-370.

Voilley N (2004). Acid-Sensing Ion Channels (ASICs): New targets for the analgesic effects of non-steroid anti-inflammatory drugs (NSAIDs). Current Drug Targets - *Inflammation & Allergy,* 3: 71-79.

Whittle BA (1964). The use of changes in capillary permeability in mice to distinguish between narcotic and non-narcotic analgesics. *Br. J. Pharmacol.Chemother,* 22: 246-253.

Winter CA, Risley EA, Nuss GW (1962). Carrageenan induced edema in hind paw of the rat as an assay for anti-inflammatory drugs. *Proc. Soc. Exp. Biol. Med.,* 111: 544-547.

Warren KS, Peters PA (1968). Cercariae of *Schistosoma mansoni* and plants: attempt to penetrate *Phaseolus vulgaris* and *Hedychium coronarium* produces a cercaricide. *Nature,* 217: 647-648.

Xu F, Morikawa T, Matsuda H, Ninomiya K, Yoshikawa M (2004). *J. Nat. Prod,* 67, 569—576.

Yoshikawa M, Wang T, Sugimoto S, Nakamura S, Nagatomo A, Matsuda H, Harima S, (2008). *Yakugaku Zasshi,* 128: 141-151.

Yoshikawa M, Li X, Nishida E, Nakamura S, Matsuda H, Muraoka O, Morikawa T (2008). *Chem. Pharm Bull,* 56:559-568

Yoshikawa M., Xu F, Morikawa T, Pongpiriyadacha Y, Nakamura S, Asao Y, Kumahara A, Matsuda H. (2007). *Chem. Pharm. Bull,* 55: 308- 316.

Yamahara J, Matsuda H, Yamaguchi S, Shimoda H, Murakami N, Yoshikawa M (1995). Pharmacological study on ginger processing: I. Antiallergic activity and cardiotonic action of gingerols and shogaols. *Nat. Med.* 49: 76-83.

Yen GH, Chen HY (1995). Antioxidant activity of various tea extracts in relation to their antimutagenicity. *J. Agric. Food Chem.* 43: 27-32.

Yoshikawa M, Yamaguchi S, Kunimi K, Matsuda H, Okuno Y, Yamahara J, Murakami N (1994). Stomachic principles in ginger. III. An anti-ulcer principle, 6-gingesulfonic acid, and three monoacyldigalactosylglycerols, gingerglycolipids A, B, and C, from Zingiberis Rhizoma originating in Taiwan. *Chem. Pharm. Bull.* 42: 1226-1230.

Yoshikawa M, Murakami T, Morikawa T, Matsuda H (1998). Absolute stereo structures of carabrane-type sesquiterpenes, curcumenone, 4S-dihydrocurcumenone, and curcarabranols A and B: Vasorelaxant activity of zedoary sesquiterpenes. *Chem. Pharm. Bull,* 46: 1186-1188.

Yoshikawa M, Sugimoto S, Nakamura S, Matsuda H (2008). Medicinal flowers. XXII. Structures of chakasaponins V and VI, chakanoside I, and chakaflavonoside A from flower buds of Chinese tea plant (*Camellia sinensis*). *Chem. Pharm. Bull,* 56: 1297-1303.

Zeng YW, Zhao JL, Peng YH (2008). A comparative study on the free radical scavenging activities of some fresh flowers in southern China. LWT.*Food Sci. Technol.* 41: 1586-1591

Zhang Y, Morikawa T, Nakamura S., Ninomiya K, Matsuda H, Muraoka O, Yoshikawa M, (2007). *Heterocycles,* 71: 1565—1576.